Averting Disaster: Science for Peace in a Perilous Age

The Erice International Seminars on Nuclear War and Planetary Emergencies: From 1981 to 2008

Averting Disaster: Science for Peace in a Perilous Age

The Erice International Seminars on Nuclear War and Planetary Emergencies: From 1981 to 2008

William A Barletta

Massachusetts Institute of Technology, USA

Henning Wegener

World Federation of Scientists
Permanent Monitoring Panel on Information Security, Switzerland

World Scientific

NEW JERSEY • LONDON • SINGAPORE • BEIJING • SHANGHAI • HONG KONG • TAIPEI • CHENNAI

Published by

World Scientific Publishing Co. Pte. Ltd.
5 Toh Tuck Link, Singapore 596224
USA office: 27 Warren Street, Suite 401-402, Hackensack, NJ 07601
UK office: 57 Shelton Street, Covent Garden, London WC2H 9HE

British Library Cataloguing-in-Publication Data
A catalogue record for this book is available from the British Library.

Mt. Cofano as seen from the San Domenico Hall, Erice, 23 August 2005
Cover photo courtesy of William A Barletta

AVERTING DISASTER: SCIENCE FOR PEACE IN A PERILOUS AGE
The Erice International Seminars on Nuclear War and Planetary Emergencies: From 1981–2008

ISBN-13 978-981-4289-40-5 (pbk)
ISBN-10 981-4289-40-X (pbk)

Printed in Singapore by World Scientific Printers

Foreword

The life of a scientist is an exciting venture. Its challenges are multi-manifold. Subject to the rigors of method, committed to the search for verifiable truth, demanding the most one can give in terms of creativity and patient labour, it is a life permanently hovering between constant frustration and occasional triumph. The scientist is also continuously placed under the moral imperative of social responsibility.

His endeavours are not limited to the work place or to the laboratory or to solitary confinement. The scientific life also brings the rewards of teaching the younger, and thus sharing insight and wisdom. It derives huge benefits from the cooperation and dialogue with colleagues that produce synergies and heuristic ideas. It flourishes in an interdisciplinary environment where a broader world view is generated that functions as a necessary reality check.

I have had the privilege of experiencing the benefits of life in a scientific community in the many stages of my career: at Columbia, at CERN, at the Institute for High Energy Physics in China, and elsewhere. But none of these holds as cherished a place in my heart as the Erice community. The unique interaction of science and government in its most constructive form, the clear orientation towards peace and global well-being, the earnestness of scientific pursuit, the harmony of purpose, the human experience, make Erice a truly exceptional place.

As a former Vice President of the World Federation of Scientists, an active participant in the World Laboratory, a proud recipient of the Erice Peace Prize, and as Co-Chairman, jointly with our much missed friend Kai Siegbahn, of the International Seminars during many years, I feel closely attached to the Erice experience. It thus gives me great satisfaction that an attempt is now made to preserve the memory of the Seminars and many of their key participants, and to assess their impact on the world at large. I am sure that this book will serve not only to keep the experience alive and duly commemorate the great scientists who have contributed, but will also put the scientific and political accomplishments and the historical importance of Erice in the perspective they deserve. May it also foster the spirit of friendship and common purpose among those who, like me, proudly belong to the Erice community and feel committed to the Erice Spirit.

It is a particular joy that this book is published at an important moment in the life of Antonino Zichichi and is a tribute to his far-reaching vision and unrelenting efforts that have created and maintained the Erice universe. For me, this is an opportunity to thank Nino most sincerely for his generosity and enduring friendship.

Tsung Dao Lee

Contents

INTRODUCTION

Introduction

William A. Barletta and Henning Wegener

This book describes one of the most exciting intellectual and political ventures of the later part of the 20th century, - of the decades of the Cold War, a period of bristling East-West tension with the omnipresent possibility that its management might get out of hand, and a war of unimaginable proportion and potential damage erupt. The International Erice Seminars, - first on Nuclear War and, in later years, on other Planetary Emergencies were the attempt, from the idyllic setting of a small location in Western Sicily, to stem these perilous tides, and to put science – "scientific culture" - at the service of political problem-solving in a new, open and interdisciplinary manner, at the highest scientific level.

Managing International Security in the shadow of the nuclear threat by a systematic exploration of the scientific, technical and cultural bases for peace was the first and foremost theme of the Seminars, but the inquiry was not limited to the search for means to avert nuclear disaster. From the beginning, and more significantly when the perspective of a massive nuclear conflict finally receded (though without disappearing), the focus of the collective scientific endeavour was on *all* major challenges to our global well-being.

The reader of this book will witness the exemplary assumption and earnest exercise of responsibility of the involved scientists in dealing with the perils of the modern age, the fragility of peace and our intensely interconnected societies, an effort sustained over well-nigh thirty years.

Hundreds of top level scientists, the best in their respective fields, among them many Nobel Prize laureates – those who had won the prize before coming to Erice, and the even greater number of those who went on to win it – have participated in the Seminars and their interdisciplinary working groups, the Permanent Monitoring Panels, and have helped to generate a steady and influential flow of scientific insight, united in their work in a unique community of knowledge, wisdom and shared values.

The book is, however, not a historical or political analysis or narrative of the Seminars contents, or their worldwide political and scientific consequences. It is a book of personal record and recollection. As many, if not most, of the scientific luminaries who formed the Erice universe in its early years have now disappeared, the time has come to commit to memory not only their role, impact and contribution to the long-term catalytic effects that Erice has produced, but also the human imprint that their presence, often over many years, has made. Their life span was shorter than the task required, and honouring their memory befits those who have benefited so much from their presence and who are still able to report about a lived experience. This brief introduction will therefore be followed by a collection of "testimonials", personal, often intimate recollections of Seminar participants.

The book is also one of gratitude. Professor Antonino Zichichi, founder and chairman of the Seminars since their inception, has also been the beacon and propelling force of the work accomplished year after year. No member of the Erice community has ever left a Seminar session without an intense feeling of gratitude to him, convinced that the relentless adherence to scientific rigor and, equally, the unique atmosphere of common purpose and comradeship so characteristic of the Seminars are owed to him more than to anybody else. No reader should thus be surprised that Professor Zichichi is the person most often cited in the subsequent contributions.

Professor Zichichi has also been the driving force in the creation of a comprehensive framework of learning around Erice. In admiration for his Sicilian fellow countryman Ettore Majorana, a genius of physics who died young, disappearing mysteriously in 1938, Zichichi established in CERN, Geneva in 1962 The Ettore Majorana Centre for Scientific Culture (EMCSC), based in Erice, Sicily. From 1963 on, university courses and advanced workshops have been organised in Erice on a broad range of sciences, attended by nearly 100'000 scientists the world over. Today, 122 "schools" are devoted to international scientific collaboration - high level teaching and debate without the pressures of examinations or diplomas. From this fertile ground originated in 1981 the International Seminars on Nuclear War. The first phase already is memorable as it saw, at the second meeting in 1982, the adoption of the Erice Statement (reprinted as Annex A), signed by Paul A.M. Dirac, Piotr Kapitza and Zichichi, a fervent plea for openness in research and science – in opposition to secret military laboratories – and for reasonable and stable nuclear disarmament.

The Statement has since then been signed by more than 10.000 scientists and a great number of world leaders. Scientific collaboration without secrets and borders, solidarity and the dedication of science to peace – often referred to as the Spirit of Erice –also gave rise to the Voluntary Scientific Service intended to share the benefits of science with developing countries. The main driving force and contributor to the Erice International Seminars is the World Federation of Scientists (WFS), established in 1973, and presided over, in eight year terms, by eminent physicists. WFS has tens of thousands of affiliates and also functions as a permanent link between scientific institutions. Through its implementing arm, the International Centre for Scientific Culture – World Laboratory (Worldlab), it has entered into agreements with Governments, scientific institutions and research centres located in developing countries. The Worldlab, created in 1986 from the midst of the WFS, promotes East-West and North-South technological and scientific collaboration through exchanges and pilot projects in accordance with the spirit of the Erice Statement. One hundred pilot projects have been implemented successfully, one hundred agreements signed with governments and scientific institutions, Forty-two centres for scientific research have been established in developing countries, one thousand international scholarships have been granted to young scientists from developing countries, while national scholarship programmes have been established in 34 countries with over 5'000 scholarships already awarded.

In 1988 the Sicilian Government, responding to a proposal by the WFS, established the annual "Ettore Majorana Prize - Erice - Science for Peace" to promote the values of scientific culture worldwide and to honour eminent persons of Culture and Science. Many recipients of the Prize are among the contributors to this book. To complete this institutional overview, reference must also be made to the many international pilot projects and influential publications that are the work products of the Permanent Monitoring Panels established for each Planetary Emergency. Professor Zichichi – and there is no more authoritative author than he – has summed up these developments in two recent publications to which the reader may be referred[1]. The institutional framework is a complex universe, of institutions intertwined, but with separate identities, yet administered jointly by an efficient secretariat in Geneva, and guided by identical values, and the same *spiritus rector* – Antonino Zichichi. From Professor Zichichi's account also emerges a sense of continuity – both in persons linked in close professional partnership, and in themes – between the early years at CERN, the Centre for Scientific Culture, the World Federation of Scientists, and the International Seminars that, more pointedly, are the focus of this book.

The ideals encapsulated in the Spirit of Erice have also determined the working method in all the institutions and programs mentioned. Starting from the premise that scientific inventions and discoveries must become instruments of peace worldwide, all major meetings have been, and are, interdisciplinary, with strong political ingredients. In Zichichi's words: "not only eminent scientists but also political leaders are needed to act so that the great achievements of Science and Technology, instead of becoming terrifying devices of war, give rise to instruments of peace and progress". But all participants had, and have, to submit unfailingly to the principles of scientific work. Planning for peace must be based on realistic and verifiable affirmations: on truth. "Truth is the result of patiently and laboriously obtaining proof…amplified by discussions, in an objective search for factual knowledge, free from assertions which are technically or scientifically unfounded." The application of these principles suffers no exceptions, with the result that even the hottest controversial topics of the era of East-West confrontation could be discussed calmly and soberly, among "free and civilized people" and with clearly demonstrated – and indeed existing - human and scholarly respect. Even in matters of nuclear strategy, nuclear war and nuclear peace, there was "no we against them" but rather a common harmony of purpose. Indeed, in the controversies directly touching on the East-West divide the atmosphere reigning at the Erice debates was the most fruitful, and often precursive of mutual understanding and peaceful outcomes. We will return to these consequences, and some concrete examples of positive impact will become evident throughout this book.

But one series of questions can already legitimately be asked at this juncture: is Erice really unique? Have the International Seminars made a quantifiable difference in the development of the crucial East-West security relationship, especially in nuclear matters? Has the Spirit of Erice measurably helped to foster peace? Have the Erice assemblies risen above the level of immensely pleasant and enriching scientific congresses at top level, as force multipliers, to be sure, of scientific insight? This book attempts to offer a carefully worded affirmative response to all these questions.

During the height of the Cold War, the international conference circus went on unabated, and multiple were the opportunities also for scientists from antagonistic systems to meet (even though those from Communist systems were carefully screened). There was the all-encompassing UN world, international scientific societies, and national initiatives. In the international security world proper, there were SIPRI, the IISS, the Institute of East-West Security Studies, the Bilderberg Conference, and Pugwash . . . a long list indeed. What sets

[1] Neither Science Nor Civilization Could Exist Without Memory. The Science for Peace Erice Prize. Antonino Zichichi. Compositori Industrie Grafiche Bologna – October 2007. This publication also contains a list of the Planetary Emergencies that are the subject of the International Seminars, classified under the following headings: Water, Soil, Food, Energy, Pollution, Limits of Development, Climatic Changes, Global Monitoring, New Military Threats, Science and Technology for Developing Countries, the Problem of Organ Substitutes, Infectious Diseases, Cultural Pollution, Common Defenses against Cosmic Objects, Huge Military Investments. Ettore Majorana. His Genius and Long-Lasting Legacy. Antonino Zichichi. Società Italiana di Fisica, 2006

Erice apart, is the rigor of science, the lack of prejudice or preferred philosophies (as so clearly and avowedly prevailed at Pugwash and, to some extent, at SIPRI) – and surely the quality of the attendees. Other distinctive features are no doubt the early opening of the Erice community to top scientific and political personnel from China (which provided some useful relativity to the direct East-West focus), and beyond that the globality of approach, as expressed in the focus on Planetary Emergencies and the growing participation by top scientists of the developing world.

Managing International Security in the Nuclear Age

From the inception of the International Seminar series in 1981, the dominant themes have been the prevention of war and the perils of the nuclear age. The early sessions demonstrate this dominance with particular force. The annual volumes of the Proceedings – splendid reading even after, or perhaps *especially* after, so many years – reflect this at times almost exclusive emphasis, and the broad spread of relevant sub-themes. They also reflect the excitement nuclear issues so understandably caused at the time. The Seminars were aptly predicated, by their very title, upon the threat of a nuclear encounter, and commenced at a time of great preoccupation world-wide. Nuclear fears ran high, the debate in the West, accompanying the Reagan presidency, was heated between nuclear abolitionists and those who saw a "present danger", the nuclear arms race was perceived to be at its height, East-West tensions foreboding. The time was ripe for bringing the instruments of science to bear on the challenge of war prevention, in a totally independent environment. Professor Zichichi has cited Italian President Pertini and Pope John Paul II as being among the many who urged him on. For the founders of the Seminars, the problem as such was not new. For Professor Zichichi and the eminent group of his colleagues, the nuclear issue – in its ineluctable ambivalence, nuclear energy being the potential benefactor of mankind and its potential destroyer – was at the centre of their professional and moral preoccupation, and had been so throughout their time at CERN, the Ettore Majorana Centre and elsewhere.

But it appeared that there was now also a shared East-West concern, and that it was possible to establish an open scientific exchange including highly placed representatives from the Soviet Union. It is one of the great achievements of the Seminars that they cashed in on this opportunity, seeking out the underlying rationality and ingrained restraint present in the leadership of both East and West. Soon, top defence thinkers, strategists and leaders – including the Science Advisors to the US and Soviet heads of State – gathered in Erice to discuss openly all aspects of the nuclear dilemma, far away from the public turmoil, diplomacy noises and polemics of official fora. The catalytic effects of the Seminars on statecraft on both sides in this era can hardly be overestimated, although the causal links cannot always be traced with exactitude.

During these initial years, there was an intense interest in the consequences of a nuclear encounter; and the apocalyptic, scary aspects of the nuclear balance and its precarious stability, predicated upon "deterrence", were clearly recognized in Erice, and dealt with from the perspective of the scientists' responsibility across system borders. What strikes the erstwhile witnesses of those early debates is both the breadth and level of discourse, and their sobriety and moderation. In the present context, only the main themes can be mentioned. A large chunk of the debates was dedicated to the consequences of nuclear war (fall-out, biological effects, pollution, climate responses, psychological effects, strategies of mitigation of nuclear war impacts). These were recurrently reviewed, particularly in the 1983-85 period. Strategies for the prevention of war, planning for peace, civil defence, discussions of nuclear non-first use, etc. completed the topics immediately relating to nuclear disaster, and new findings or theorems were reported on an annual basis.

The Erice community was never anti-nuclear in an emotional sense. Realism, perhaps even a basic consensus on the indispensability of nuclear weapons and their stabilizing potential in a particularly perilous configuration of world politics prevailed, and formed the underlying fabric for the debates. Their focus was on harnessing the nuclear threat by mustering all means of science and technology to make its management rational and peace-prone. Arms reduction, incisive nuclear disarmament and the quest for strategic stability, providing a "controlled context", achieving "reasonable sufficiency", a rational relationship of nuclear and non-nuclear armament were concepts discussed with knowledge and vigour, - and often without tangible East-West controversy.

During the early and middle eighties, three problem areas stood out before others: the relationship between nuclear offence and defence, the climatic consequences of a nuclear exchange under the "nuclear winter" hypothesis, and the emerging spirit of East-West cooperation in the Gorbachev era.

The thinking of the leading minds at Erice had always nourished a positive prejudice in favour of nuclear defence, both from the technological and the moral angle. President Reagan's March 1983 Strategic Defence Initiative, proposing to "intercept and destroy strategic ballistic missiles before they reach our own soil" and that of the Allies, thus propelled the debates on strategic defence into high gear. Erice had the benefit of having some of the best and most influential actors in the defence technological field right on stage: Edward Teller, Lowell Wood, Evgeny Velikhov. The debates, which covered a good deal of the 1983 to 1985 Seminars, were both fascinating and portentous of consequences. Perhaps the most fascinating was the nearness of views between the US and Soviet representatives, evidenced by a 1983 project at Erice to "establish a joint USA-USSR and European research group" to undertake the simulation and evaluation of the global consequences of a USA-USSR nuclear confrontation, but also – more important in this context – to study "a new type of defence system against nuclear destruction". Clearly, there was a shared view that science could clarify both the technical possibilities and the strategic advantages to be gained from defence, and the endorsement of that document by scientists occupying high positions in their respective administrations testified to that conviction. Probably, nowhere else were SDI and the fundamental defence-offence relationship been discussed with such intensity, depth of knowledge, technological sophistication, - and shared political goodwill.

The nuclear winter scare of the early eighties had enhanced wide-spread nuclear fears in a way difficult to imagine now that the topic has simply gone away. For the Erice community, in-depth consideration was attractive, because here two of the two major Erice preoccupations, the earnest study of the consequences of nuclear disaster, and the concern for world climate, coincided. The debates on the possible global and stratospheric effects of a major nuclear onslaught were thus wide-ranging and allowed for excellent insights in the then state-of-the-art of modelling climate developments; especially noteworthy in this respect were the contributions by the late Professor Joseph B. Knox from the Lawrence Livermore National Laboratory. Yet, a clear-cut response to the nuclear winter hypothesis did not emerge: scientists differed considerably as to the probability of consequences, their intensity, duration, geographical implications, etc., a telling indication that the sophistication of climate modelling had not advanced to the level that characterizes today's debates on global warming and climate change.

As from 1986, the advent of Gorbachev's policies and the new opportunities for arms control made their imprint upon the agenda and the tone of the Erice Seminars. The atmosphere at Erice had always been one of sober exchange, a scientific community spirit and mutual understanding, but the arrival of Soviet scientists directly involved and influential in the changes in their country, improved the atmosphere further. With the imminent prospects of the INF Treaty to be concluded, the START negotiations in process, and substantial progress in the Geneva negotiations on a Chemical Weapons ban, the debate on nuclear matters now concentrated on disarmament and its consequences in terms of strategic stability and new force configurations. The realignment of the earlier power structures – regional stability patterns, European security structures, NATO enlargement, and new battlefield configurations - was debated under the new basically cooperative setup. Post-disarmament policies were also discussed: what to do with retired missiles (INF, ICBM, warheads), for scientific peace projects and economic conversion. Decommissioning and its consequences, including decontamination and stockpile disposal, remained a subject for several years, - indeed until 1998.

As the spectre of nuclear all-out war disappeared, the Seminars could revert to their broader agenda of science for peace and the safeguarding of the planet. The major concern shifted from "nuclear war" to "nuclear peace".

Yet, the years of a more cooperative stance in the relations between the West and the (former) Soviet Union after the fall of the Berlin Wall did not obliterate the prominence of the nuclear and armament agenda. The new key concepts for discussion were – and still are – nuclear non-proliferation, a shared concern of all, and proliferation of other weapons of mass destruction and - of course - of missiles. Nuclear defences continued – and continue - to be investigated, in the nineties under the aspect of cooperative models, as of 2007 again under a more antagonistic paradigm (the "Bush-Putin disagreement"). Global stability through disarmament obviously remained high on the agenda in all these times. In 2008, non-proliferation was a subject of central interest to which a whole plenary day was dedicated.

But there was an important shift of emphasis in the nineties and the early 21st century. Beyond the East-West perspective, the international security debate necessarily became multi-polar, global. And, even before September of 2001, international terrorism became the concept *à la mode* that crystallized attention. The conceptual framework included – here as elsewhere – "asymmetrical threats", new ways of warfare and conflict that called for new strategies of conflict resolution and their scientific underpinning. Nuclear, chemical and biological threats were examined under these new circumstances ("nuclear terrorism", "bioterrorism"), nuclear, chemical and biological safety appeared among the new challenges. Under the Planetary Emergencies paradigm, the "cultural emergency" called for the scientific analysis of international terrorism in all its aspects. Outside of Energy and Climate, perhaps the most important focus of the Seminars in the past few years has been on Terrorism, from its root causes, its difficult to fathom psychology, motivation, to mitigation strategies and countermeasures. Special attention has also been given in this context, in the framework of in-depth work on digital security, to the growing threat of cyber terrorism and cyberwar. This work, in the light of the globally disruptive effect of international terrorism and its consequences for world order, will no doubt reflect a major focus of the work of Erice also in the future, proving the topicality and necessity to probe, at all times, the role of science and technology in the generation of peace.

The Role of Science in Managing Planetary Emergencies

While the potential for nuclear miscalculation and consequent holocaust was pre-eminent in the early years of the Seminars, it did not completely dominate the attention of the participants, nor was the connexion of the nuclear technology and wider issues of human welfare ignored. The closest scientific connexions were the consequences of human activity on the global climate and nuclear energy to fuel global development. Both issues have persisted to the present.

However, as recalled in the preceding section, by the late '80s the spectre of nuclear winter had disappeared as a concern, both because of the emergence of a more fruitful relationship between the East and West driven by the policies of Mikhail Gorbachev – and, prompted, the Teller disciples would have said by the SDI – but also due to the recognition that the alarmist claims by the Sagan camp were based upon crude and deficient climate models. By the time the Berlin Wall had fallen, experts from Lawrence Livermore National Laboratory, who had had such an influence on earlier nuclear winter debates at Erice, shifted the concern of the assembly to the potential of the increasing emission of greenhouse gases to generate significant warming of the climate. Perhaps, deriving from the high standards concerning numerical modelling held by Zichichi and the other high-energy physicists at the Seminars, the early debates about global warming focused on both the accuracy and evidential basis of climate models. Those concerns have persisted to this day.

As early as 1993, the contrast between the proper scientific scepticism and quasi-religious alarmism had become stark in the debates. At that time, following a data-filled presentation, the *"opposite type of talk"* was given by a member of the Stockholm Institute (SIPRI). The presentation was filled with half-truths and with phrases intended to get an audience to infer something that could not be supported from the facts alone. When he was caught on a few of these types of intellectual dishonesties, his standard reply was, *"I didn't say that."*. The exchange illustrated the dangers of politicized science as well as the necessity for scientists to engage in the education of the public and the policy-makers. This dedication toward rigorous examination and scientific scepticism has been a hallmark of the Erice Seminars.

More recently the WFS Permanent Monitoring Panel on Climatology summarized the status of the scientific inquiry and public debate:

> What was once a relatively easy and low-cost task of obtaining data for studying and predicting these changes, is now becoming expensive, complicated and threatening as data are copyrighted and offered on a 'for sale' basis by international coordinating bodies. Global monitoring of trends requires inter-comparability and continuity of key observations, combined with the recovery of historical information. Unfortunately, observation systems for gathering climatic data are becoming increasingly costly and difficult to maintain. Furthermore, some of the standard systems upon which climate research depends (e.g. the international upper-air sounding system) are being eroded. The quality of the information provided to the lay public, industry and governments is critical to the public

> perception of this issue and the scientists studying it … Unfortunately, the quality and reliability of the information is highly variable and is sometimes distorted.

One approach to ameliorating adverse human effects on the environment is through what is now called geo-engineering. One early example was a proposal to remediate ozone depletion in the upper atmosphere by electrolytically precipitating out the chlorofluorocarbons. The precipitators would be carried on giant lighter-than-air balloons. The audience at Erice found many holes in the proposal; nonetheless fifteen years later, scientists talk about open air capture of CO_2. Carbon capture from smokestacks and subsequent geologic sequestration have brighter prospects and continue to be examined by joint sessions of the Energy and Climatology PMPs within the framework of the August Seminars. Yet a third approach, examined in 2006, proposed changing the albedo of the north polar regions by injecting appropriate substances, high in the atmosphere. While many "greens" find a moral hazard in geo-engineering – carbon sequestration mostly excepted – it is to the credit of the Nino Zichichi's International Seminars that such proposals can be examined openly and dispassionately. Open science must rise or fall on its own merits.

The second scientific theme – advocating a return to an expanding role for nuclear energy to fuel the global economy in rich as well as developing nations – was a favourite emphasis of Edward Teller and of Antonino Zichichi. Nino was especially interested in providing a venue for physicists of the highest calibre to convince the public that Italy, poor in native energy resources, should re-embrace nuclear power. In a manner typical of the International Seminars, the question of enhancing safety was met head-on. The Chernobyl disaster, the flaws in reactor design, the gross misconduct by the operators and the hazardous securing of the reactor were examined in depth at a time when fears of fallout contaminated food were still rampant in Italy.

In later sessions, the international sharing of nuclear technology has been but one example of the aforementioned thrust – alas now spent – toward the civilian conversion of military assets. As a member of the NATO Science Committee, Zichichi offered the framework of the 18th International Seminar to the "NATO Workshop on Global Stability through Disarmament" that was organized by his long-time friend and colleague, Thomas Ypsilantis. As part of the program, advanced reactor designs such as molten salt, thorium cycle systems with pyro-processing were discussed, as well as more radical proposals such as Teller's idea of building buried, semi-autonomous reactors with strongly negative reactivity coefficients. Richard Garwin recalls:

> Teller maintains that reactors and people do not belong together, so that reactors should be placed 100 m underground and only communicated with via regulation of the coolant flow. The reactors should be buried without possibility of access, and it is to be hoped that they would have a core of 30-year life after which they would end their operation and entomb themselves.
>
> There should be no moving parts in the reactor, and it should be foolproof, because there should not even be fools involved. Teller says "Don't use dangerous fast breeders, which use a metal coolant that is flammable. Use a thermal breeder, using Th232. Then end of life will not be due to lack of fuel but probably to components giving out." The resultant core will be highly radioactive and self protecting because of the U232 produced by (n-2n) on U233, which after a number of alpha decays ends in Tl208, which is a potent gamma emitter.

Nor were the issues of non-proliferation, waste disposal, and elimination of the world's plutonium ignored. Garwin noted that the chances of a nuclear explosion in a city were greater then than during the Cold War. The issue of plutonium and the nuclear fuel cycle has remained on the agenda to the present. The debate over once-through, open cycles versus closed cycles is as open now as it was fifteen years ago and is regularly revisited by the Permanent Monitoring Panel on Energy. Energy is indeed a Planetary Emergency that requires constant and incisive attention.

In such situations replete with international complications, some people may wonder what a small group of independent scientists can do. It is, however, just where one finds a confluence of national and commercial pressures, that an interdisciplinary, international group of scientists can have a critical influence on the public opinion as well as on government action. Nino addressed directly this question in 2002:

> For those of you who participate in our Meetings for the first time, and who might think we are 'poets and dreamers', let me tell you why we believe we can contribute to the solution of what has become the most important of all Planetary Emergencies. We are the largest international 'no strings attached' scientific community of volunteers. We can call on thousands of scientists the world over, who all achieved the highest responsibility in advanced research activities, covering all fields of Science and Technology. Let me recall that twenty-two years of activity in project implementation and the results represent our best source of self-confidence.

Already five years earlier he had reassured us:

> Fifty projects have been implemented for the benefit of fifty developing countries, in practically all fields of science. The World Laboratory action is very much appreciated by the beneficiary countries, for its independence from political pressures, the high degree of motivation of its volunteer scientists, its pragmatic but very strict professional approach and its sound management of project funds. The leverage effect provided by the voluntary commitment of scientists, the free availability of well-equipped laboratories, the contributions made by the governments of developing countries, its low level bureaucracy and the freedom of its equipment-buying policy, has increased the productivity of the World Laboratory projects three-fold, when compared to other similar international projects.

The International Seminars and the World Federation of Scientists have emphasized that the continuing environmental pollution of the Earth and the degradation of its ecosystems constitute an overwhelming and encompassing Planetary Emergency, that requires the greatest possible international East-West and North-South cooperation to implement effective ongoing remedies. Early discussions of the destruction of the large stores of chemical weapons in the International Seminars led to the establishment of the International School of Innovative Technologies to Clean the Environment, which presents regular sessions to this day. The topics for that School continue to be stimulated and vetted by sessions at the International Seminars. Among the more recent topics examined is the widespread dispersal in groundwater of methyl tertiary-butyl ether (MTBE), a compound widely used as a gasoline additive to improve combustion efficiency and reduce emissions of air pollutants. Initial laboratory tests indicate MTBE is a carcinogen. The increase in environmental contamination associated with organo-chlorine endocrine-disruptor chemicals (EDCs) is becoming a global problem, resulting in serious negative consequences for both human and ecological health.

While the consequences of dioxin exposure have been widely discussed in the public press, the effects of minute quantities of the nearly omnipresent plasticizers is poorly known. The International Seminars and the Pollution PMP provide a high visibility venue to address this global issue.

Although the International Seminars have been the brainchild of physicists, the critical issues of human health have not been neglected. Early on, the mechanisms of AIDS and its epidemiology have been examined. Concrete projects such as studying the wildfire spread of AIDS in Africa have been regularly reported as have the Worldlab efforts to provide some relief to the victims of this disease on the sub-continent. A concern of the past ten years has been to monitor the new biological and genetic engineering technologies: the application of genomics and proteomics, and especially the imminent possibilities of man-made, self-organizing and self-reproducing systems. Monitoring would serve to: identify the ethical and political issues arising from biotechnology; provide an early warning system for potential dangers and abuses; identify balancing tests to weigh potential benefits against dangers and, finally, recommend guidelines and regulatory measures if necessary.

The challenge to the scientists, to dedicate themselves to the solution of Planetary Emergencies and problems of global scale have always been at the forefront in the Erice International Seminars. The threats are faced by all peoples and all nations, rich and poor – the burden on the poor being especially harsh as it attacks human dignity. The Spirit of Erice with respect to the obligation of industrialized nations – and especially on the scientific endeavour in those nations – is best expressed in the Farnesina Statement, written in November 1988 by the Co-chairs of the International Seminars: Antonino Zichichi, Tsung Dao Lee, and Kai M. B. Siegbahn.

> We believe that one of the most serious problems facing the World today is *Poverty and the Loss of Human Dignity*. A large part of the world's population is in poverty due to overpopulation, famine and other natural and man-made causes. With extreme poverty, there invariably comes a loss of human dignity, which indeed has an equally damaging effect.
>
> With the recent relaxation of East-West relations, the threat of nuclear war may have been reduced. The lack of human rights still exists, and remains a major problem. But the deprivation of human rights comes from without, while the loss of human dignity occurs from within. Human rights can readily be returned if the external oppressive force is removed. However, human dignity, once lost, may not be regained. Consequently, to tackle the human dignity problem, food and medical aid is far from sufficient.
>
> We are convinced that a direct way to combat the human dignity problem is to help the people in the Developing Countries to acquire self-confidence. The most effective method is through the transfer of Science and Technology. Only when scientists from the Third World can contribute to Science in a meaningful way, would they and the people there be truly self-confident. We should, therefore, restore human dignity through Science.

The teaching is not one-way. Rather East, West, North and South can learn from each other. T. D. Lee, a great exponent of this point of view offered us a lesson in describing how developing countries might contribute to solving problems in the industrialized world. His example was the possible use for the Mississippi and in Missouri of a flood control system similar to that developed, in 250 BC, for the Mijang River in China. In fact, such a system of a by-pass canal for heavy rains plus an irrigation system is now used in the US near Wichita, Kansas.

Nino Zichichi often has reminded us of the distinction between science – knowing the nature of the physical world – and other great forms of human intellectual accomplishment – mathematics, language, and art. In his essay *"Language, Logic, and Science"* first presented at Erice, Zichichi offered his perspective on the progression from language to logic to mathematics to science as first practiced by its progenitor Galileo Galilei. To be sure the essay itself was not science but rather a combination of history (Science started with Galilei 400 years ago), epistemology (Economics is not Science), metaphysics (Immanent Reality is based on 3 pillars), aesthetics (The only "boundary condition" of art is to be magnificent), and sociology (No fellow can be called a Physicist if he has never….). Yet, delving into the nature of the order and integration of scientific knowledge is essential to understanding the roles and boundaries of science and the moral responsibilities of scientists *qua* scientist. As a contrapuntal illustration of the exuberant combination of knowing and creating, Zichichi has throughout the years offered the participants of the Seminars examples of Science connecting with Art. Despite their direct focus on concrete problems and global human emergencies, the International Seminars transcend technocracy and mere problem solving. By making an expansive call to the very best intellects on the planet, the International Seminars represent a profoundly humane example and exploration of what the existentialist philosophers have called being-in-the-world, that is, participatory ties of humanity to this world. (In analogy with sport, it is a view of the game from the field not from the stadium.) Taken in context, the Seminars – especially those concerning the most threatening existential issues confronting the planet – have touched on the authentic in Being, the beautiful in Creating, the good in Doing, and the true in Knowing.

A vexing question that was confronted many times at Erice was, in the language of the popular press, the "responsibility of the scientists". To be sure, knowing *per se* is morally neutral, yet the act of knowing involves – in addition to the formal object (the what) – both instrumentality (the how) and intention (the why). The technological axis that connects knowing and creating requires doing, controlling Nature. Crossing the moral sphere in moving from doing to knowing implies moral responsibility. Therefore, on the platform of P.M.S. Blackett Hall (the former San Domenico lecture hall) many of the most illustrious scientists at once accepted personal moral obligation while focusing moral responsibility on political leaders.

Despite the focus of the International Seminars on the liberating potential of Science – knowing evaluated against experimental and observational data rather than personality and authority – the overwhelming impression of first time participants, and even on many "old hands", is that offered by the intellectual giants that have occupied the centre of the stage at Erice. Scientists like Paul Dirac, Eugene Wigner, Piotr Kapitza, T.D. Lee, Kai Siegbahn – to mention only some of Erice's Nobel Laureates – Edward Teller, Evgeny Velikhov, Richard Garwin, Richard Wilson and our host Antonino Zichichi drove the dynamics of the Seminars. They spoke frequently about a wide range of topics even when as Edward Teller told us, "I don't know. But I will tell you." Their example, and especially that of Antonino Zichichi, have inspired the rest of us to make the International Seminars on Nuclear War and Planetary Emergencies a transformative human experience.

This Introduction cannot be concluded without gratefully acknowledging the contribution of the staff of the International Seminars and of the Secretary General, Dr. Claude Manoli.

Without the efficient and generous help from this team – untiring, infinitely helpful, precise, often invisible, but always present - the Seminars, and the entire Erice operation both in Geneva and on location, would have been unthinkable in all these years.

This book would also not have seen the light without the invaluable collaboration of Claude Manoli.

TESTIMONIALS

Andrei Andries

The creation of the World Federation of Scientists (WFS) by a group of eminent scientists led by Isidor Isaac Rabi and Antonino Zichichi in the period when mankind lived in the atmosphere of the Cold War, was an unprecedented step for the unification of more than 10'000 scientists against the danger of a planet-wide nuclear disaster. This wonderful initiative contributed significantly to the end of the Cold War. Another milestone in the activities of the World Federation of Scientists, under the leadership of Prof. Antonino Zichichi, was the identification of 15 classes of Planetary Emergencies and the foundation of the International Center for Scientific Culture, the ICSC-World Laboratory, in Geneva for the better organization of the fight against the threats worldwide. It was my honor and pleasure to participate in several Seminars organized by the ICSC-World Laboratory in Erice and Rome, and I was impressed by the wide range of problems discussed at those seminars and by the very high professionalism of discussions.

It is of great importance that the Federation promotes international collaboration in science and technology among scientists and researchers from all parts of the world. It is hard to overestimate the contribution of the WFS in the support of young scientists in new independent countries.

As is well known, at the end of the last century in some countries in the transition stage, scientific activity was considered an element of secondary importance in the hierarchy of political, social and economical priorities. This resulted in many negative consequences such as: the reduction of financial support by the state for research and technological activities, the restriction of receptivity of economical and industrial structures for scientific-technological products and applications, the underrating of scientific work and the falling of prestige of science in society, the exodus of researchers (brain-drain) into other areas of activity, to other countries, deterioration of technical and material base of research, material, social and moral decline of research staff. It is true that, now, the authorities of Moldova have undertaken some measures for improving the situation in science. Parliament has approved the Code on Science and Innovations, remuneration of scientists has been improved and the government has delegated its functions of management in science and innovation to the Academy of Sciences of Moldova. Nevertheless, much more effort should be undertaken in order for the situation in science to be considered satisfactory, especially in attracting young people into science.

Now the Doctoral studies in Moldova take 3 years (4 years for part-time learning) but a regular monthly scholarship is very low.

There are of course many international programs and grants to support Moldovan Young Scientists. Among them there are the Moldovan Research and Development Association – MRDA (Branch of US CRDF); the German Academic Exchange Service (DAAD); EU Programs, NATO Science Committee programs; scholarships offered by the French Government and several embassies, for example. The sums of these grants are large, but at the same time the expenditures of the projects are large as well. This is because conditions of grants require researchers to travel to respective countries, and this is covered by the grant, as well as certain living expenditures. This is why the proposal of Prof. Antonino Zichichi, made in 1999, on behalf of the WFS, was very attractive for us because students could continue their studies in their 'native' conditions.

Of course our main concern is not only to select the best postgraduate students but to be sure that the contents of their projects satisfy strategic directions of the WFS in terms of Planetary Emergencies. To our great surprise many fields of research of Moldovan scientists and scholars are very close or even coincide with the priorities of the WFS.

At present, the strategic directions of activity in science and innovations in Moldova are as follows:

- Establish the rule of law and the value of the cultural and historical heritage of Moldova in the context of European integration.
- Use the human, natural and informational resources for a sustainable development.
- Biomedicine, pharmaceutics, health care and strengthening.
- Agricultural biotechnologies, soil fertility and food safety.
- Nanotechnologies, industrial engineering, new products and materials.
- Efficiency of the energetic complex and ensuring the energetic security via the use of the renewable resources.

Nowadays, according to these strategic directions in the Academy of Sciences, fundamental and applied research are conducted in the following fields of science: problems of mathematics and informatics; theoretical physics; solid – state physics; micro- and optoelectronics; transfer processes in electric and magnetic fields; geological and geophysical processes; physical and technical problems of power engineering; chemistry of coordinated compounds; bioinorganic chemistry; physiology and biochemistry; microbiology; ecology; protection, renewal and rational utilization of flora and fauna; physiology of stress; genetics, medicine; history; linguistics and literature; philosophy; ethnography; art; economy, etc.

So, to a great extent, the strategic directions of science in the Republic of Moldova approved by our parliament, as well as specific areas of fundamental and applied research carried out at the Academy of Sciences of Moldova coincide with several important priorities of the WFS, and therefore are a good basis for solving the problems in conformity with the priorities of the WFS.

The Academy of Sciences of Moldova is also an important national center for scientific post-graduate studies in 17 fields of science. This includes 75 specialties, the training of highly skilled scientific personnel who work productively in scientific research institutes and in institutes of higher education, the economical sphere, the financial – banking sphere and in central and local public administration bodies.

Since 1999 the WFS has awarded 77 research scholarships. Now our people can do investigations without leaving the country (including 2008/2009 academic year) in the field of basic research in mathematics, exact, technical, humanitarian and economic sciences. It is very important that Moldovan young scientists under the supervision of the best scientists of

Moldova have been involved in the investigations of important priorities of the WFS such as the study of water problems, food and energy safety. Many projects were oriented at solving ecological problems, problems of pollution, analyzing climate changes, global monitoring of the planet. Great attention was paid to the problems of medicine and biotechnology, measures against diseases, as well as the problems of new materials and new technologies. Some aspects of social problems in the Republic of Moldova, limits of development and cultural solutions have been studied too.

We consider that this is a good example of the fruitful cooperation of the WFS with the Republic of Moldova, which began 10 years ago. This is why we greatly appreciate the contribution of the WFS and personally its President Prof. Antonino Zichichi in the financial support of young scientists from developing countries of the former Soviet Union, Moldova being one of them. The scientific community of Moldova expresses gratitude to Prof. Antonino Zichichi, the President of the WFS, as well as to all WFS staff for their noble contribution in preserving and developing the scientific potential of the Republic of Moldova.

William A. Barletta

August 19 through 24 – like the feast days of the saints, the dates are immutable and the venue venerable. Yet in the City of Science and Peace, the ceremonies and celebrations of Erice International Seminars have been worldly in focus and universal in intent congratulatory messages from the Vatican notwithstanding. Through the more than twenty-five years of the Seminars, the presiding eminences have included illustrious scientists – Nobel laureates, advisors to heads of government, heads of national academies of science – diplomats, and political leaders of Italy. The extraordinary motive force of Seminars has been Italy's most publicly recognized physicist, Prof. Antonino Zichichi, his goal to focus the attention of scientific leaders upon issues of existential importance and by so doing to alert and energize both the public and its elected officials.

First impressions (1985 and 1986)

At their inception in 1981 the Seminars carried the ominous title, Seminars on Nuclear War. Not surprisingly the delegations from the United States and the Soviet Union, led by the iconic figures of Edward Teller and Eugenji Velikhov respectively, were the largest and most vocal. Their respective nations held the largest nuclear arsenals and thereby gave the greatest urgency to find ways to diminish the chances of a nuclear exchange.

One line of debate that dominated the earliest seminars was the potential for man-made disruption of the global climate. At that time the specter was nuclear winter, which soon morphed into acid rain, only to emerge twenty years later as global warming. By my first visit to Erice in 1985, the debate over nuclear winter had been overshadowed and revealed as a peripheral warm-up engagement between East and West. President Ronald Reagan had announced the Strategic Defense Initiative (SDI); Edward Teller had been anointed by the Western press as its high priest. The time had come for a direct confrontation of the scientists. Velikhov v. Teller; Western Democracy v. the Evil Empire. On the left side of the hall of San Domenico facing the Dirac Equation, the Soviets and the East; on the right side Prof. Zichichi (and in later years his co-chairs,T.D. Lee and Kai Siegbahn), Edward Teller and the West. (An exception to this arrangement was Chinese Ambassador Quian Jiadong seated next to T. D. Lee.)

Our trip to Erice was a mini-drama that foreshadowed the session. Arriving in Rome from San Francisco, my family and I were greeted at the stairs of the TWA DC-10 by a smartly dressed Alitalia hostess and taken to a special VIP lounge for passport formalities and to wait for our connection to Palermo. In Palermo the scene was repeated; a black Mercedes on the tarmac, special assistance in retrieving our baggage, and then the hour-long ride toward Erice. Leaving the highway near Val d'Erice, I became aware of many police sirens. What was happening? A look around revealed the situation. Two vans of carabinieri in the lead followed by a procession of polished vehicles; we were the first car. Once we arrived in from of the Majorana Center (San Rocco), I moved to exit the car, but the driver instructed us to wait. Soon the back door of the van in front of us opened, carabinieri leapt out with their assault rifles and we were motioned to exit the car.

Even as an employee of the US nuclear weapons lab at Livermore, I was not used to security like this. The International Seminar was no ordinary scientific conference. The first session was to feature an extended address by Hon. Giulio Andreotti, Foreign Minister of Italy.

Tensions between Italy and Qaddafi's Libyan government were high. No wonder security was tight and top carabinieri brass were present.

The debate over strategic defense followed three main themes: 1) Technically SDI would or wouldn't work – Teller and Lowell Wood led the US charge on that front; 2) Even if the defensive forces work technically, they would destabilize the strategic balance between the US and USSR; this line of debate was waged most vigorously several years later by Teller disciple Greg Canavan against Gorbachev advisor, the mathematician Andrei Piontkovsky. 3) The US and USSR would be better to work on technology jointly; this was the line championed by Velikhov and exemplified by his request to put networked personal computers in that hands of Russian school children. Moreover, Velikhov echoed the contrapuntal theme strongly and frequently advanced by Zichichi in his addresses to the assembly: Open Science, without secrecy, without borders, North, South, East and West. The broader strategic context was left to the NATO allies to expound, most notably by Ambassador Henning Wegener.

To be sure, other luminaries graced the stage in the early years: Nobelists Eugene Wigner, speaking quietly in conciliatory tones; Sir John Eccles, wary of the Russian bear, Sam C. C. Ting, focused on international scientific cooperation; Kai M. B. Siegbahn, carrying a gentle, humanistic message; T. D. Lee, offering incisive analysis of the primary theme of the opening session. In each case our host, Prof. Zichichi, pointedly questioned the speakers knowing that the overall message of the Seminars would need to be melded into a coherent whole.

Then as now, Italian press prefers high drama and personality over cold hard facts; they swarmed around the major actors of the great Star Wars debate wondering, "what does this all mean? What does this mean for Italy? Are the scientists getting the public into a new existential crisis?" The major part of the interaction with the press was in collective interviews after the morning sessions. The rapporteur to the public of the proceedings was our host. Always a master of the art, Nino managed and directed these meetings with the press – his principal message that science (and scientists) offered the opportunities to defuse the threats of mutually assured destruction and to address the other human emergencies of planetary scale. The message from Erice was not only carried in the press, but because of Zichchi's close ties to the Italian government to the Italian Foreign Ministry.

Velikhov's arguments frightened Teller; they exploited Teller's oft-proclaimed belief in scientific openness. Yet they came from the scientific advisor to the leader of the USSR, the land that Teller's patron had labeled the Evil Empire. Unlike at many scientific conferences where Russians tended to keep to themselves, at Erice they shared song and drink late into the night with their Western counterparts. Was this a seductive trap? Was it just a clever joke when Velikhov explained that the Russian ballad being sung translated as, "The boy and the girl stand in the field watching the missiles flying peacefully toward the United States?"

Within several years Chairman Gorbachev would dissolve the Soviet Union. The threat of mutual assured destruction would be replaced by the threat of Soviet nuclear materials falling into the wrong hands. Star Wars would be widely recognized as the most intellectually debilitating exercises visited upon the US scientists and technologists. At the International Seminars fifteen years later when one scientist after another (including many Americans) decried the counter-terrorism actions of the US, their lone strong defense was urgently offered by Andrei Piontkovsky. Erice had provided a unique venue for scientists to analyze and influence world events.

A post-Chernobyl episode

In 1988 the Seminars on Nuclear War explored a new nuclear threat, managing the consequences of the Chernobyl reactor disaster. Evgenji Velikhov being absent, the Russian delegation was headed by Acad. V. D. Pismenny, Director of the Kurchatov Institute at Troisk. Pismenny headed the technical aspects of securing the damaged Chernobyl reactor core, receiving a substantial radiation dose himself. By now the policy of glasnost was in full bloom and the behavior of the Russians was markedly changed. The presentation of technical details of the Chernobyl disaster presented in great detail, and at every meal I was button-holed by Pismenny and several of his colleagues to join them.

Coming from a nuclear weapons lab, I had always been cautioned about being pressured for information. Nothing of the sort! Indeed, I could not get in a word edgewise. One after the other, my Russian companions told me about environmental incidents and accidents at their labs – about reckless disregard for public and worker safety. One of them told me about an enormous gas pipeline explosion that I was to learn nearly twenty years later was actually caused by a CIA act of cyber-sabotage.

At Livermore, I had studied the Russian activities in high current induction accelerators, both betatrons and radial line linacs. US Air Force intelligence had proclaimed these devices to be strong evidence of a Russian Star Wars program. I thought otherwise. One afternoon I asked about the machines, technical questions why were parameters chosen. My Russian companions explained that these accelerators were designed for the nuclear weapons program in the USSR, with parameters selected to meet flash radiography requirements. Details that I knew from the literature were expanded upon and explained without prodding.

As soon as one explanation was offered another topic was begun. It was the chance of a lifetime for these scientists to speak freely about their work and about incidents that disturbed them. For me it was like a priest hearing confessions.

As every night after dinner there was singing, marsala and vodka, that is until the last night before departures. That year Prof. Zichichi had arranged for our excursion to visit the Mondo X enclave on the small island of Formica in the Bay of Trapani. On Formica we saw the restored tonnara, handiwork of the recovering drug users that Mondo X was founded Padre Eligio to serve; we swam in the cool, crystal clear water; we were served a simple but elegant meal before departing accompanied by the President of the Region of Sicily. Once back in Erice, we were taken to the Municipio for a greeting by the Mayor and for each to receive a memento, hand-painted plate with the seal of the town of Erice and a personalized inscription on the back.

As we were leaving the Muncipio after midnight, Pismenny called to me and Eric Storm – head of the laser fusion program at Livermore, "We need to talk." We walked back to his room in San Rocco with three of his most senior associates. "Tonight no vodka." Pismeny told us about the Soviet inertial fusion program. We talked about what areas of international cooperation in fusion were active. He then told us that he had a proposal for US-Soviet cooperation that had been cleared "at the highest level of the government." He asked if we would informally bring this information to attention of the US Department of Energy (DOE). If there was any interest he would come with a formal proposal when he visited the US the following October. The Soviets did not want to make a formal proposal only to be promptly and brusquely turned down. He explained in great detail. At 4:45 am he announced, "Okay, you have to go. The driver comes in 30 minutes to take us to the airport and we haven't packed."

This was the quintessential Erice experience. Eric brought back the proposal much to the consternation of the DOE and the State Department. LLNL was rebuked. Scientists should not be talking about such things. That was the quintessential bureaucratic experience back home.

Catching time to discuss physics with Nino during the Seminars was nearly impossible. A better prospect was an afternoon encounter on the beach at San Vito lo Capo during the week between the International Seminars and School of Sub-nuclear Physics.

From 1986, I had been working on power sources for a TeV scale linear collider of electrons and positrons. I knew that Nino had been advocating the Eloisatron, a machine five times larger than the Superconducting Supercollider (SSC) under construction in the United States. One afternoon at San Vito, I asked him to consider staring to support the linear collider. He responded with the claim that the most exciting physics required a greater energy reach, that we did not really know how to build a linear collider, and that one could certainly build a machine much larger than the SSC. I responded that I doubted the latter claim. Whereupon Nino challenged me to prove my claim by doing the analysis and by organizing a workshop to consider the question, "what is the highest energy proton collider that could be built. I agreed to the challenge. That conversation led over the next several years to a series of Eloisatron workshops, two books, and eventually the Very Large Hadron Collider (VLHC) effort in the US.

Having discussed physics, Nino asked me to help him get his windsurfer. As we were walking down the beach, a well-tanned fellow in his early twenties came running up to us. "Professor Zichichi, Professor Zichichi." Nino shook his hand. "I want to introduce you to my girl friend," the fellow responded as the young woman approached. Nino smiled. I was amazed. In the US I could not imagine any 20 year-old guy recognizing any physicist, let alone wanting to introduce his girlfriend.

I had come to understand the impression that Nino had made on the Italian public about the importance of science. Some years later as the 400th birthday of Galileo approached, I asked Nino why he was making such a public acclaim about Galileo Galilei. He replied, "I want the Italian people to understand that Galilei is as much a part of their patrimony as Michelangelo and daVinci." Nino echoed the urgency of educating the public about science sometime later at a meeting of the World Federation of Scientists, "You will know that you have done enough when in the back of fashion magazines, the pages about astronomy are at least as many as those about astrology."

An historic session

A recollection about the East meeting West at Erice would be incomplete without describing the memorable 1991 session. In 1990 Academician and basso profondo Evgenij Shelkov and others introduced the "planetary emergency" of the threat from cosmic objects. Consequently the 1991 Seminars had a special session attached concerning space technology that attracted an unusually large (more than 20) delegation of Soviet scientists and technologists. Even more unusual was that many of the Soviet delegation were in attendance with their wives. The head of delegation Acad. Velikhov had with him his wife and his daughter, Natalia. The Velikhovs and most of their delegation arrived on August 18.

On August 19 we awoke to press reports of the decree of Gennady Yanayev announcing himself as acting President of the USSR due the "illness" of Mikhail Gorbachev. Within hours Boris Yeltsin appeared at the White House, home of the Russian parliament, and proclaimed the decree to be a reactionary coup attempt. The actual condition of Gorbachev was unknown. For three days, especially in the evenings, our Soviet colleagues were drawn to the two televisions in the Madonna Room – the set on one side tuned to Russian State television; the set on the other tuned to CNN. Typically they watched Russian TV for a few minutes, then moved to the other side to learn what was really happening. I recall standing next to Mrs. Velikhova watching the Russian news feed on the second evening. She muttered an expletive in English under her breath and then tuned around to view CNN.

During those three days several of us asked Eugenij whether he would return to Russia. He had been candid about the danger to his life if the coup were sustained. Yet his answer to this direct question was always the same, "I will return if President Gorbachev asks me to return." Later I learned from Lowell Wood that a military driver was on alert to drive Velikhov and his family to the US Naval Air Station at Sigonella (near Catania) where a US military plane would transport him to asylum in the US.

Throughout those three days the marsala room was atypically silent after dinner. Happily by the fourth day the coup had collapsed. That evening an emotional impromptu celebration was in order.

Random recollections

- A surprise morning excursion from the military airport at Trapani with Sir John Eccles and the Italian Minister for European Affairs to meet Zichichi, Sam Ting, and Velikhov for an audience with Foreign Minister Andreotti with the aim of promoting a recommitment to nuclear energy in Italy.

- Lowell Wood volunteering to receive the world's first brain prosthesis.

- Edward Teller's scandalized response to a Los Alamos talk about Accelerator Transmutation of Waste; "There is no waste; there is only treasure!"

- An address in Italian by Nino to an elegantly dressed assembly of the Rotary Club of Trapani. The topic, the Nature of Space and Time; the audience was spellbound.

- First meetings with Kai M. B. Siegbahn that eventually led to my becoming his successor as Coordinating Editor of Nuclear Instruments and Methods – A.

- Watching my daughter playing soccer on the beach at San Vito with Nino's three sons and their friends. Girls playing soccer just wasn't done then; not in Sicily.

For all these experiences and many more I offer my hearty thanks to Antonino Zichichi – genius, guide, and generating force of the International Seminars.

A sacred place

For nearly three millennia,
perched upon an isolated mount overlooking hills and plains –
resplendent now with figs of India,
fields of melons, yellow skinned with mint-white flesh,
and vines of Bacchus closely cropped –
The temple of the golden goddess
received the supplicant and wayfarer.
From there (some say) next to the stars,
the Cyclops hurled the Farallones
that threatened brave Ulysses.
From that ancient womb a magic city, dedicated to her, grew -
a center of the Magna Grecia.
Her temple now in ruins,
the priestesses of sacred rite dispersed,
Still the goddess beckons searching spirits
filled with passion for more worldly wisdom.
Entering the ancient walls,
once celebrated by the poet of the birth of Rome,
the modern pilgrims still may be transformed.
their charge to sing the joys and pains of her devotions.
Some songs are evocations,
Some songs are introspections,
Each song an exploration
of the labyrinths and convolutions
of encounters archetypal
with Venus, Vesta, and the Muse.

Victor Baryakhtar
Valery Kukhar
Gennady Palshin

Ukraine. 1986. Chernobyl. The explosion of the reactor at the nuclear power plant (ChNNP) – the largest technogenic catastrophe in world history.

Right here in Ukraine, in the course of the Chernobyl accident mitigation, the principles of the Erice Statement found a wide practice ground. Here Professor Zichichi's strikingly piercing vision of cooperation between scientists of the World without frontiers and borders for the good of the whole human race has taken true shape.

Let us remind ourselves that the 4th unit of ChNNP contained approximately 231.5 tons of nuclear fuel; 20 to 50 % of volatile radio-nuclides released in the air exposed of the population in the wide areas of Belorussia, Ukraine, and Russia was and resulted in the radioactive contamination of enormous area. In Ukraine alone, 73 administrative districts, 2294 settlements and 3.5 million people were exposed to contamination.

For a few days, elapsing from the moment of failure, radio-active clouds ran across the European territory of Russia, the Urals and further towards Siberia. Some contaminant streams reached Scandinavia, the Balkans and Turkey, leaving tracks in most countries of Europe.

Certainly, with his insight in the nature of the global planetary emergencies, Professor Zichichi could not but come to the very focus of the accident – the site of the nuclear power plant. In the immediate vicinity of the destroyed reactor, in front of nuclear power plant scientific staff and the press, he formulated, both the basic principles to follow in developing the nuclear-generation industry as well as specific tasks - for scientists in working out the operation of technologies for safe renewable energy, and mass media in strengthening public awareness of their safety.

The branch of the International Center for Scientific Culture «World Laboratory», established in Ukraine as the result of Professor Antonino Zichichi's efforts, and for many years coordinated international research activities focused on Chernobyl-related issues. Understanding the great potential of this collaboration, the National Academy of Science of Ukraine and ICSC-World Laboratory agreed on an international scientific center, «Institute of Chernobyl» to be founded at the Ukrainian Branch of the World Laboratory. This center, under the leadership of Professor Baryakhtar, covered research in a wide range of issues – from the state of the destroyed reactor up to the temporary shelter «Sarcophagus», and on to on-site contamination, exclusion zone, ground and surface waters, migration of radionuclides etc.

For over 20 years the seminars on Planetary Emergencies in Erice have provided a venue for Ukrainian scientists to share their research achievements and recommendations on reducing the negative impact of the radioactive contamination on human health and the environment.
Among those who reported on Chernobyl-related research findings from the Erice podium are two vice-presidents of the National Academy of Sciences of Ukraine – Professor Kukhar and Professor Baryakhtar, the Minister of Public Health of Ukraine – Professor Serdyuk, the Head of the State Committee for Science and Technology – Professor Ryabchenko as well as the

renown Ukrainian scientists – geochemist Professor Sobotovich, biologist Professor Mikhailov, Gennady Palshin, Director of the World Laboratory, Ukraine, and many others.

No one in the Erice audience could remain unconcerned about the subject of Chernobyl - questions, remarks and suggestions voiced here have been inspired by huge intellectual potential, creative thinking, uncommon scientific insight of Professor Zichichi, Edward Teller, T.D. Lee, Kai Siegban and other outstanding scientists of the Erice "circle" enriching Ukrainian scientists with a new vision and ideas.

Fixed in my memory is a point which once evoked the special interest of Professor Zichichi: «How to distinguish the impact of radionuclide exposure on worsened health indicators in contaminated areas from other factors (stress, overall environment pollution etc.)?» - things taken for granted and obvious at first glance did not look so indisputable on the Erice stage. Not only did this questions compel us to give our assumptions new consideration, but it also gave impetus to search for new ideas and research methods.

For example, the discussions gave start to new research activities –such as the Retrospective Dosimetry Method by Tooth Enamel, which arose following a report made in Erice by Professor V. Baryakhtar. This project funded by the ICSC-World Laboratory ensured that Ukrainian specialists had reliable retrospective dosimetry tools, and was of great importance from both a medical and social viewpoint – since the individual exposure rate is taken as a basis for benefits defined to the exposed people.

On the whole, the Ukrainian Branch of ICSC-«World Laboratory» carried out dozens of scientific research projects on Chernobyl-related issues, which involved scientists from Ukraine, Russia, Belorussia, USA, France, Italy, Germany etc.

The non-governmental status of the World Laboratory provided a favorable background for the idea of inter-agency teams, which united the scientists from the National Academy of Science, the Ministry of Public Health and those for Chernobyl Issues, specialists in the area of agriculture, and higher education faculty.

Research programs implemented within the framework of the World Laboratory Institute of Chernobyl under the leadership of Professor V. Baryakhtar covered in particular:

- «Antropogen» covering issues of the health of the population exposed to anthropogenic accidents;

- «Immunocor» dealing with research of immune response in the population in a contaminated area;

- «Radlan» focused on radio-nuclide mapping of the territory of Ukraine;

- «Sarcophagus» aimed at the development of the best solution ensuring «Shelter» safety for the environment

Large-scale generalization of the above research projects, which took place during the Erice seminars, ensured the cross-disciplinary review and systematic approach to effects generated in the wake of the Chernobyl accident, which consecutively brought about a set of concepts implemented later in government regulations, i.e.

- Concept of safe residence in areas exposed to radioactive contamination;
- Concept of the «Shelter» transformation in the environmentally safe system;
- Concept of radioactive waste treatment in the nuclear generating industry of Ukraine;
- Complex concept of the ChNPP Exclusion Area.

The fact alone of Erice seminar participants visiting the Chernobyl Nuclear Power Plant, to meet with scientists and specialists working there and to communicate with people living in the exclusion area, clearly states the interest in this theme on their part. Worthy of special note are the visits of Professor I. Ortalli and Professor Geobel, whose study of safety of the destroyed unit at the very beginning of that work was of great significance.

Professor Zichichi's fruitful efforts in facilitating the dialog between scientists of the West and East and the systematic approach to the Planetary Emergencies classification and study, were highly valued by Ukrainian scientific and by the political elite - the President of Ukraine, Leonid Kravchuk, welcomed Professor Zichichi during his visit to Kiev, while the National Academy of Sciences awarded him the status of Honorable Member of the Academy for outstanding contributions to the development of Physical Science and worldwide scientific progress.

Juan Manuel Borthagaray

My first meeting with Professor Zichichi came during my first term as Dean of the Faculty of Architecture (FADU) of the University of Buenos Aires. We completed a successful period of cooperation with the Department of Architecture (DA) of the Swiss Federal Polytechnic at Lausanne (EPFL) and had agreed to seek the cooperation of the ICSC-World Laboratory for the second phase. That was my first glimpse at Professor Zichichi's lion-like, white chevelure. Our Buenos Aires Center for Computer Assisted Design at FADU UBA was born there and then. Afterwards, I had the opportunity to meet Professor Zichichi personally, and several meetings followed, both at Lausanne and Geneva. But it wasn't until after my first Erice experience that I really came to appreciate Nino. Erice meant, and means conviviality, the chance to share a table with the Chinese Ambassador to the Disarmament Conference at Geneva, from whom I learned the true meaning of the word aristocracy of the best kind. What could an urbanist and architect have to do among such hevyweights of science and cold war politics? Then the organization showed that its title "International Centre for a scientific *culture*" and its proclamation of interdisciplinarity were not *pour la galerie*. A paper of mine "Metropolis, the most dangerous place in earth" was welcome, and so the meetings opened a window out from physics, proliferation and A bombs, and I was encouraged to invite, a Brazilian (Serra) a Mexican (Pozo) and an American (Ellis), colleagues to form a Permanent Monitoring Panel (PMP).

Meetings on air and water pollution in big cities, solid waste disposal and migrations followed, as well as dinners at delightful Erice *trattorias* with Ambassador Ahmad Kamal, plenary sessions with the thundering Oxford accent of Sivaramakrishnan, and on a more intimate scale but no less thundering, requests from him to San Rocco's waiter for yet another, very little, glass of *lemoncello*. Some of my best recollections are of the departed astrophysicist Mayo Greenberg and CERN physicist Douglas Morrison. So many Asians, but also Europeans, like dear Hiltmar Schubert and our perhaps best Erice friend, Bertil Galland. Furthermore, the American scholars who have moulded culture through the 12+ years of Erice meetings attendance, and whom we would never have met if not for Nino's wonderful initiative. Now, a word from my professional innermost self. What a wonder Sicilian Territory and landscapes are. What a wonder Palermo, the ruins of Segesta and the exquisite beauty of Erice, with the traces of all the cultures that left their imprints in layer upon layer. Every year, when I enter San Rocco and see the cortile, and the breakfast room I have that strong feeling of being home at last. Home is also the beautifully austere, welcoming cells at San Francesco, where corridors, cloister and vaults are a masterpiece of restoration. Notwithstanding all this praise, I would without any doubt award the palms to the architect of San Domenico, with the grand aula for the General Assembly and, above all, the breathtaking coffee break terrace, with the monumental view to the hills and the Mediterranean.

On a final note. I was amused, recently, to read a comment in the New York Times, on the ICSC, at the time of the 2008's Sessions, in which the reporter noted that Professor Zichichi ruled everything uncontestedly, from the sessions programme and the seating arrangements to the choice of menus and dinner wines. I wrote to Secretary General Claude Manoli about the note, and he told me it was from a very well known scientific reporter, invited by the Centre. In the times of the old Romans, when the Republic was in danger, it was statuted that the Senate could appoint a dictator for a very brief term, not to outlive the republic's emergency. For my part I must say, that if Nino is to be dictator, may my enological emergency last for ever.

Enzo Boschi

Although I am more familiar with the School of Geophysics, the most prestigious school and advanced workshop on Solid Earth physics in the world, founded by my teacher Professor Zichichi, I have regularly attended the International Seminars on Nuclear War and Planetary Emergencies from their very beginning.

Erice is an enchanted place, born again, after centuries of neglect, thanks to the stubborn efficiency of Nino Zichichi who, sparing no effort managed to have it rebuilt as the City of Science. During the last 46 years in this primordial city sung by the Greeks, Nino managed to weave a web of advanced schools, workshops and seminars in most domains of Science and human civilization which, still today, are unequaled in the world. More than a hundred schools and workshops are held every year, dealing with frontiers of science topics, under the leadership of the most eminent scientists of the planet.

The crowning achievement is the International Seminars series on Planetary Emergencies. These Seminars are reserved for a multidisciplinary audience of eminent scientists, and each session is dedicated to the latest progress achieved for a certain Planetary Emergency, which are then debated in a multidisciplinary manner. Scientists vie for the honor of presenting their latest findings in front of such an audience, capable of detecting flaws in related domains and suggesting new avenues in research. Because of my early association with Erice, I was lucky to meet and discuss with some of the most legendary scientists who helped shape our science in the last century: Weinberg, Abdus Salam, Glashow, 't Hooft, Charpak, Ting, Lee and dozens of others.

Erice has served Science and scientists, all over the world, for the last 46 years, in a very unique and exceptional manner. Tens of thousands have witnessed it first hand but still too many actors, in our circle of decision-makers, have not yet realized the full importance of its planetary impact. It is quite unprecedented in human history that achievements of this magnitude can be attributed to a single man, who kept it going for half a century. I therefore state it out, loud and clear, Professor Antonino Zichichi, who should have received, a long time ago, a Nobel Prize in Physics for the excellence of his scientific research, certainly deserves a second Nobel award, the one for Peace.

Vladimir B. Britkov

"The only real luxury - it is the luxury of human communication"
Antoine de Saint-Exupéry

The World Laboratory and the World Federation of Scientists, organized and led by Professor Antonino Zichichi, provide an excellent opportunity for interesting and valuable communication and contact between remarkable people, starting with Professor Zichichi himself, at meetings set up in a wonderful, unique place - Erice.

For example, in 2008 I met and established friendly contact with a high-caliber professional and historical personality – the pioneer of the internet – Professor Robert Kahn. Professionals I manage to meet on a regular basis; however, to meet such a person who has entered into history was certainly exceptional. An amusing story is related to that meeting:

In 1994 in Moscow I met a certain Dr. Alexander Kahn, who was introduced to me as Robert Kahn's brother. He was starting the Russian edition of PC Magazine and needed a welcoming article that I wrote, together with Academician Stanislav Emelyanov. Meeting with Robert Kahn I told him that I had had the pleasure to be acquainted with his brother. Robert Kahn replied that he was very pleased to hear this, but that he had no brother.

Working for many years in the PMP on Information Security I had the opportunity to communicate not only with colleagues in my narrow specialty of IT System Analysis, but with very interesting people with different backgrounds, such as the chairman of the PMP on Information Security - Ambassador Henning Wegener (Germany) and the chairman of the PMP on Terrorism, Ambassador Ahmad Kamal (Pakistan + UN). Such contacts expand one's professional horizons. Another example is the contact I had with an expert in the field of climate from Michigan University, Natasha Andronova, who participated in the report of Albert Gore, a Nobel Prize laureate. It turned out that Natasha Andronova worked in Leningrad (Saint-Petersburg) before 19 August 1991 (putsch and the collapse of the Soviet Union).

Thus the World Laboratory allows for the meeting of those Russians whose paths would not necessarily cross in every day life, for example, Arthur Chilingarov - corresponding member of the Russian Academy of Sciences, Soviet Hero and famous polar researcher.

The interdisciplinary approach developed by Professor Zichichi provides an assurance that all participants (members of WSF) will find new approaches to the problems addressed.

Franco M. Buonaguro

To fully understand the role of the Erice Seminars and Professor Zichichi in non-mathematical-related fields, it is relevant to analyze his scientific approach to bio-medical subjects. At the beginning of the Western culture, great intellects (often philosophers) were involved in all fields of human knowledge and were applying similar epistemological methods in scientific as well as non-scientific fields, with different results in terms of quality and relevance. Paradigm of this condition is Aristotle, who focused his interest in all fields of knowledge and, besides being the teacher of Alexander the Great, elaborated the first taxonomic classification of natural beings and the first encyclopedia. Later philosophers became more and more specialized in Metaphysical versus Physical subjects, and within the physical area (particularly in the Modern culture since Leonardo da Vinci) a further differentiation in Chemists, Physicists, Biologists, Physicians, etc was introduced. However, also in contemporary periods multidisciplinary approaches are needed and this is especially true in highly complex systems. People, able to bridge different disciplines, although not frequent, represent the exceptions who can accomplish high achievements. My first two astonishing friends were the Italian Physician/physicist Salvatore Luria [Nobel Prize Awardee in Virology, 1969], who moved to radiation biology in Rome and worked with Fermi; and the American Civil Engineer Mortimer M Elkind [Enrico Fermi Awardee, 1996], who moved to radiation biology and identified the toxicity and geno-toxicity of X-ray radiations at Argonne National Laboratory and Fermi Laboratory in Chicago. This was my personal experience when I met Nino Zichichi and I was introduced to his multi-task approach.

Zichichi's Schools and Centers

My first personal contact with Nino was in 1982. I was just back from Elkind's laboratory at Argonne National Laboratory (ANL) in Chicago. During my stay at ANL Mort deeply introduced me to the beauty of the calculus and the need of rigorous application of statistical and mathematical methods in bio-medical research. I was looking for a larger international context for my cultural growth and I ended-up at the Majorana Center in Erice. The course, directed by Claudio Nicolini[1], a bio-physicist involved in DNA chromatin and genotoxicity studies, was focused on Chemical Carcinogenesis. The unconventional approach was evident since the first lectures. Furthermore, during my stay in Erice I had the opportunity of getting acquainted with all different schools based at the Majorana's Center. The following year (1983) I moved back to the US with a World Health Organization (WHO) fellowship to be spent, at the Fred Hutchinson Cancer Research Center, in the Virology Unit on the role of Herpesviruses on Kaposi's Sarcoma, whose epidemic was already attracting the interest of the scientific community[2]. While the World was discovering the devastating reality of the AIDS epidemic and was looking for possible strategy to contain its exponential transmission in developing countries, following the African AIDS Conferences we organized, with several colleagues (in particular Gaetano Giraldo and Elke Beth-Giraldo), our research proposal, and Zichichi included in the World laboratory program the establishment of two African AIDS

---

[1] Nicolini C. **Chemical Carcinogenesis**, *NATO ASI Series A: Life Science Series Vol. 52*, Plenum Press, New York and London, 1-489, 1982.

[2] Conference on Kaposi's sarcoma epidemic, Washington 1983

Research Centers (AARC): The West AARC located in Nigeria and the EAST AARC located in Uganda.

The East Africa AIDS Research Center (the first AIDS Research Center in Uganda)

The East Africa AIDS Research Center was established at the Uganda Virus Research Institute (UVRI) in Entebbe in 1995. The broad scientific approach and the profound humanity of Nino, allowed, through the World Laboratory, the restructuring of the allocated laboratory rooms and their furnishing with appropriate scientific instruments for the isolation and molecular characterization of HIV field isolates. Several exchange missions were supported for Italian scientists visiting the EAACR as well as for Ugandan colleagues visiting the Naples Cancer Research Institute. Furthermore Workshops and Conferences were organized and held in Uganda to implement clinical diagnostic procedures, particularly for blood bank safety, and biomolecular analyses inviting several clinicians and scientists from adjacent countries.

Meanwhile, within the context of the Erice Seminars, several sessions were organized to bring together top Western scientists and Developing Countries scientists/physicians, to have the most updated view on the epidemic and the scientific discoveries to be quickly translated to the field.

Biomedical Research and the mutidisciplinarity

Since 1998, a further development of the Erice International Seminars has been the establishment of the World Federation of Scientists Permanent Monitoring Panels, one of which focused on AIDS and related cancers. This panel, in the subsequent 10 years, organized annual workshops and plenary sessions on specific aspects of the AIDS epidemic, including Mother-to-child transmission and anti-HIV vaccine approaches. The plenary sessions of the Erice Planetary Emergencies Seminars were the most peculiar component. They were planned on the basic Zichichi concept: the multidisciplinarity. The practical implication of this unusual model was that in the same conference room (along with several nobel laureates) physicists, biologists, physicians, chemists, epidemiologists, veterinaries, climatologists, science writers, philosophers, politicians, and scholars of the majority of disciplines were discussing on the same subjects, tackled from all different view points.

In the 2008 the Planetary Emergencies Seminars focused on Climate changes, the global temperature rise has been analysed in respect to several scientific aspects. It would not be unusual to start a session on the new energy fonts, to lower the emission of greenhouse gas, and to end up with Public health planning to contain emerging epidemic of vector-borne diseases, in unusual geographic regions, because of temperature modifications on the globe. In previous conferences, focused on environmental contaminations, it would have been commonplace to discuss nuclear waste, nuclear accidents of several megaton bombs and the analysis of polar bear's teeth for monitoring environmental contamination. A multiplicity of critical subjects discussed in a multidisciplinary environment with global experts on specific components willing to listen and to contribute to other scientific fields. Between over a hundred scientists, in the first row, Nino was always present, ready to request further information, to identify some incorrect analyses, to understand the scientific problem and contribute to its solution.

Galilei's role in Modern Science

Galilei's methodology goes beyond the physics limits and represents also the fundaments for most biomedical disciplines. Very often, particularly in philosophy, the empiricists activity and the following illuminating studies are referred to Bacon's inductive methodology for scientific inquiry, often called the *Baconian method* or simply, the scientific method, described in the Novum Organum[3], in contraposition to Aristotle's Organum. In those years, Galilei was conducting his studies, establishing a revolutionary scientific approach and collecting most of the results, which later on, would be used by Newton to establish some of the most relevant physics laws. The Galilei method of testing and testing again has become the key component of modern scientific approach. However, for some unknown reason, in the Anglo-Saxon culture Galilei is not taken into due consideration. Bertrand Russell, a Nobel laureate in Literature (1950)[4] and one of the most relevant philosophers of the XXI century, in his masterpiece on the Western Philosophers "A History of Western Philisophy", in a brief chapter on Galilei's achievements identifies Galilei as *the greatest of the founders of modern science*[5], in his Biography, expresses instead his preferences to Newton's mathematically oriented work over Galilei's more empirically oriented approach. Nino has been able to emphasize achievements and role of Galilei, whose work was often contrasted by the contemporary Religious culture and only partially supported by the Academy of Lincei. On the contrary, Newton achievements have been all along strenuously defended by the Royal Society (Britain's Academy of Science), of which he became the president, even against Leibnitz' claim for the development of Calculus.

Galilei showed a remarkably modern appreciation for the proper relationship between mathematics, theoretical physics, and experimental physics. According to Stephen Hawking[6], Galilei probably bears more of the responsibility for the birth of modern science than anybody else, and Albert Einstein called him the father of modern science[7].

Why bother with Galilei's scientific method, based on innovative combination of experiment and mathematics, and characterized by the willingness to change views in accordance with observations? In recent years, even in the Western Medical Community, it has been necessary to introduce the concept of Evidence-based medicine, in order to optimise disease's treatment and ban expensive ineffective therapeutic protocols. It is however in Developing Countries that there is the need for a science-based medicine. One of the major fights to contain AIDS transmission has been conducted against peoples' traditional belief in different countries and context. In sub-Sahara Africa the major routes of AIDS transmissions have been fuelled by pagan believers: to gain health by having sex with young virgins, resulting in the rapid transmission of HIV – infection from HIV-infected old rich men to young innocent girls; to clean survivors from death impurities by sexual interaction with a Mwesha (sanctifier), with consequent transmission of HIV infection from one tribe to the other. A further relevant role

---

[3] Francis Bacon: Novum Organum 1620

[4] "in recognition of his varied and significant writings in which he champions humanitarian ideals and freedom of thought", The Nobel Foundation (1950). Bertrand Russell: The Nobel Prize in Literature 1950.

[5] Bertrand Russel Chapter VI The rise of Science, in A history of Western Philosophy by Bertrand Russell, pg 531 1945

[6] Stephen Hawking. A Brief History of Time. New York, NY: Bantam Books. p.179, 1988

[7] Albert Einstein. Ideas and Opinions. translated by Sonja Bargmann. London: Crown Publishers. p. 271, 1954 "Propositions arrived at by purely logical means are completely empty as regards reality. Because Galileo realised this, and particularly because he drummed it into the scientific world, he is the father of modern physics—indeed, of modern science altogether."

in AIDS transmission has been the denial of a viral cause, even by the past-President of South Africa, which has hampered for several years the possibility to implement preventive transmission measures. Without a modern scientific attitude based on identification and characterization of causal agents, we would still be blessing or accusing God for epidemic/pandemic diseases without being able to identify and prevent pathogen-driven infectious diseases.

Zichichi's role in Bio-Medical Science

Besides the attention paid by Profesor Zichichi to scientific/public health subjects and his propensity to financially support projects on bio-medical subjects, Nino has played another major role for several scientists, including myself: the need to speak-up, the relevance of being an active testimonial of ones own discovery's and own ideas. Scientists are often not public men, and are unable to express their ideas in straightforward sentences. This has several major draw-backs: the modest interaction between scientists and public, the presence of journalists interposed between discoveries and the audience with the frequent manipulation of the main messages, to make the news more appealing for the general population. The remark on the need to speak-up, as physicists have done during the cold war to contain nuclear armament, is often initially perceived in a negative way. The need to present ones own ideas, also when not fully accepted, becomes, however, a way of life. As an example, our mobilization during the Naples waste crisis. Among the multidisciplinary subjects explored by Nino, within the activities of the WFS, a few years ago we focused on the environmental pollution (chemical, nuclear, biological). Nino, among several environmental scientists, invited also Guido Bertolaso (Director of the National Civil Defense), who described the Italian status on natural hazard (e.g. volcanic eruption, earthquake, or landslide), nuclear and house waste. In the year 2008 Naples had a major garbage collection/handling crisis, solved by the relevant role of Bertolaso in identifying the appropriate waste collection site and garbage-processing strategy. During this civilian/urban crisis several scientists (including myself), usually hesitant to speak out in public contexts more suitable for politicians, became involved in public conferences.

Concluding Remarks

It has been a great privilege for me to have the opportunity to interact with Nino Zichichi who, at scientific as well as at personal levels, has been able to enrich my personal life. Without his profound knowledge and willingness to contribute to the human progress I, along with my colleagues, would not have been able to contribute to fighting against the HIV epidemic in the most remote areas of the globe.

Gregory H. Canavan

In the depths of the cold war there were few places where it was possible, let alone acceptable, to discuss global problems and their solution. Erice provided such a venue. Professor Zichichi built it by inviting friends of international stature to visit Erice to discuss fundamental problems in science, technology, and society. Gradually the discussions were broadened to the more sensitive issues of global war and its consequences, which ranged from strategic forces and their stability to missile defenses and their impacts. Erice was one of the few places that these problems and their possible solutions could be discussed in a dispassionate and productive manner. Much of the reason these discussions remained objective and productive was Professor Zichichi's "gentle" prodding of participants towards a useful solution that all could accept.

All was not deadly serious. I often accompanied Dr. Teller to the meetings, which he enjoyed enormously because they recalled the free-wheeling discussions he participated in when quantum physics was in its infancy. It was also pleasant to see him interact with Prof. Lee, who still gave Dr. Teller the deference due his old professor, and Dr. Garwin, who had worked with Dr. Teller in Los Alamos.

By the end of the cold war Erice was recognized as a valuable site for such discussions. Perhaps for that reason, when the transfer of power in the Soviet Union evolved into an attempted coup, President Yeltsin sent a large contingent of scientists in his own plane to participate in the Erice seminar. It soon appeared that this contingent was not chosen randomly, but might contain many of the scientists who knew their missile launch codes. Despite their senior status, they quickly proved themselves to be competent scientists and enthusiastic participants.

A by product to that interaction developed the following year when the Russian economy faltered and its science needed external support lest nuclear scientists leave Russia. U.S. scientific contingents formed by State and Commerce for extended fact finding tours to Russia were hosted by many the same people who had come to Erice during the attempted coup. We found that they were the heads of critical design bureaus, but they freely discussed proposed military and civil projects. The teams negotiated projects for cooperation in science and technology that could have been of great benefit to both countries had they been accepted by U.S. administrations.

By that time there were a number of venues for discussion and a large number of groups visiting Russia, so one might think the possibility for contributions by Erice might be diminished. But most of the groups that visited Russia left little behind, particularly in the critical area of energy. That left Erice an important role for the interchange of information on the development of an energy infrastructure that respected the environment. In those discussions western scientists collaborated with Russian scientists like Academicians Velikhov and Fortov, who had often been on opposite sides of earlier debates, in developing the architecture of Russia's energy infrastructure.

Discussions have turned cold again. One can say that the chill is due to geopolitical factors or military actions, but much appears to be due to lack of communication. Discussions of strategic missiles revive the arguments of the 1960s, and discussions of missile defense ignore the insights developed in Erice in the 1980s and 1990s. Discussion has sunk to levels not seen since the depths of the cold war. In the areas of science, energy, and military much could be gained by resuming those earlier dialogues. While there are now other venues for holding such discussions, none has the scope and record Erice achieved under Professor Zichichi in its prime. It may be time for Erice to resume the unique service it performed in previous decades. It could be important for it to do so.

The author is a Senior Fellow and Science Advisor at the Los Alamos National Laboratory in Los Alamos, New Mexico, USA. The views expressed here are his own.

Georges Charpak

(in collaboration with Nathalie Charpak)

My first major experiment with CERN, where I had been hired by Leon Lederman, who was spending a sabbatical year there, was to measure the anomalous magnetic moment of the muon, in collaboration with a small group of physicists, directed by Richard Garwin. I had the privilege to meet Antonino Zichichi who was amongst the young Europeans participating in this exceptional project.

I contributed on many occasions to the seminars held at the Centre of Physics in Erice, founded by Professor Zichichi. In conversation with him, I mentioned the admiration that I had for the work undertaken by my daughter, Nathalie Charpak, paediatrician in Bogota, in the development of a method of care for premature infants called the Kangaroo Method, which had been invented by a Colombian doctor. Nathalie endeavoured to further, by scientific methods, this progress which was the result of an empirical activity.

Professor Zichichi had been impressed by this work and believed that the World Laboratory, which he had founded, could contribute to the development of the project. He proposed that I present, in three days time, a report to the Scientific Committee of the World Laboratory. This was surely the most stressing presentation that I had had to make in my career. For three days, day and night (thanks to the time difference); my daughter taught me everything that I know today on the subject of premature infants. Due to the scientific competence of the committee, composed primarily of physicists, it was evident that I had to be clear in my explanations on the Kangaroo Method. My strength of conviction, coupled with the enthusiastic support of Professor Zichichi, persuaded the committee to accept the Kangaroo Method.

My daughter could thus benefit from constant support for many years in order to tackle the scientific and social problems in the implementation of this method.

Born in Colombia in 1978, in the mind of a Colombian paediatrician (E. Rey deceased 1994), the MMK made a gigantic leap from the original Colombian idea, a local method, slightly folkloric, to a general method of care in neonatology, based on solid scientific proof. This leap was possible thanks to the support of Professor Zichichi: “the price of the rigorous evaluation of the Kangaroo Method is peanuts compared to the enormous benefits which this method could bring to premature infants of the world”. He was a visionary, even if this field was far from his area of expertise.

This method is currently experiencing world-wide development, and rarely does a congress go by on premature infants, without there being a section devoted to the progress made thanks to the scientific methods which are now used to solve the problems which appeared. Nathalie goes to Erice each year promoting tirelessly the Kangaroo Mother Method (MMK).

Thousands of premature infants of low birth weight have today benefited from this method, whether from western developed nations or third world developing nations.

More than 50 teams from 25 countries have learned the MMK at the Kangaroo Foundation of Bogota (hppt://kangaroo.javeriana.edu.co) and for the majority; this is thanks to the help from the World Laboratory which has since ten years, provided MMK training scholarships to developing countries. The World Laboratory has displayed a steadfastness and exemplary logic. Firstly in providing support for the rigorous evaluation of the method, and secondly in the dissemination of the programme to those countries which are the most needy, and which could most benefit from the method.

The WHO published in 2003 a practical guide of the Kangaroo Method to assist the implementation of this procedure in the institutions of poor countries (www.who.int/reproductive-health/publications/kmc/kmctext.pdf).

We must all fight together for appropriate solutions in the development of these countries, but with equity, so that our fragile children will not only survive, but with a quality of survival guaranteeing optimal brain development; so that we will have future citizens capable of changing and improving the world. This is something we really need!

Luisa Cifarelli

I have known Professor Antonino Zichichi since the early seventies of the last century. I met him for the first time in Erice. In 1968 my father, Michele Cifarelli, had been elected Senator in Sicily, in the electoral district of Marsala-Trapani. The beautiful city of Erice was part of it. For many years we spent our summer holidays in Erice where very soon my parents had bought a charming little house for this purpose. Erice is where my father and Zichichi had the chance to get to know each other and establish a deep and long lasting friendship. As an Italian and European politician, for during the same period of time my father was also a member of the European Parliament, he was invited to attend the very first INTERNATONAL SEMINARS ON NUCLEAR WAR, held at the "Ettore Majorana" Centre, thanks to the unprecedented initiative of Zichichi. I remember my father's utmost interest in following the Seminars, his esteem and admiration towards his friend Zichichi who had assumed the noble mission to promote, well ahead of times, the wonderful concept of "Science for Peace". In this spirit the Erice Statement was written down by P.A.M. Dirac, P. Kapitza and A. Zichichi in 1982, when the fall of Berlin Wall was simply unimaginable.

Coming back to my first encounter with Professor Zichichi in Erice, I just remember the only question he asked me (I was still an undergraduate student at the "Lycée Chateaubriand" of Rome, in the class of "Sciences Expérimentales"): *«Why don't you come and study Physics in Bologna?»*. Bologna, his beloved Bologna, was the seat of his university. This early question was the beginning of a fascinating scientific adventure with Professor Zichichi. He was indeed my thesis supervisor in 1975 when I got my "Laurea in Fisica" in Bologna. Since then I am still a member of his research group. He gave me the great and unique opportunity to participate in the most advanced experiments and projects he has been, and is still, carrying out in the major European laboratories for particle physics, such as CERN, DESY, and Gran Sasso.

I did not closely follow the INTERNATIONAL SEMINARS ON NUCLEAR WAR, which soon became the INTERNATIONAL SEMINARS ON PLANETARY EMERGENCIES. But I have been involved with all ranks (as student, lecturer, invited scientist) in the INTERNATIONAL SCHOOL OF SUBNUCLEAR PHYSICS, the glorious and worldwide renowned Erice School for post graduate physicists, which was at the origin of the foundation of the "Ettore Majorana" Centre for Scientific Culture (now also Foundation) in 1963. I also had the privilege to participate in the start up and organization of the series of INTERNATIONAL WORKSHOPS OF THE ELOISATRON (ELN) PROJECT, a series implemented by Zichichi in Erice, as from 1983, in order to promote intense R&D (research and development) studies on a future super powerful proton-proton collider.

Due to my lengthy, extensive and varied work next to Professor Zichichi, till now under his inspiring leadership, I dare to feel somehow allowed, in this testimonial in his honour, to sketch a brief portrait of him as follows.

The main characteristic of Professor Zichichi is encompassment, in very many sectors of human knowledge and human acitivities.

During his long scientific life, Zichichi has been able to make forefront research on fundamental aspects of subnuclear physics, spanning from the discovery of the first example of nuclear antimatter, to the first meaningful search for a new sequential heavy lepton, to the search for new phenomena such as the appearance of particles originating from the Superworld at an unexpected energy scale or emerging from the Color-Gluon Coloured World that could be probed according to him in a totally unconventional way. His scientific encompassment includes a large assortment of ideas, of course, but also the invention and implementation of very many challenging instruments and techniques capable to achieve his goals. His scientific encompassment moreover comprises an amazing vision about the evolution of scientific research which has always been out of the common run. One example for all: he has launched the first worldwide project of a super proton synchrotron collider of extreme energy and luminosity, called ELN (as mentioned above), largely predating the idea of the present LHC collider – a 10% scale model of the ELN – that will finally come into operation at CERN this year. And LHC could be built at CERN in the same tunnel that previously housed the LEP facility thanks to his formidable suggestion in 1979, which he put forward notwithstanding the common wisdom of the time, to build such a tunnel with a much larger cross-section than initially foreseen. In addition to all these achievements, he has been able to open his mind to interdisciplinary problems for the benefit of mankind, rigorously addressing numerous planetary emergencies that seriously threaten future generations.

Turning to people and society, in relation with science and culture, Zichichi has been able to create and develop laboratories and education centres, such as the INFN Gran Sasso underground laboratory and the "Ettore Majorana" Centre in Italy. He has shown on several occasions his capability to fight at the decision making level to obtain the right funding and welfare level for research and researchers. He has always been able to motivate young people by establishing many new fellowships, prizes, grants, in a high spirit of generosity and indulgence. He has been vigorously standing up, as said, well before the fall of the Berlin Wall, to trigger and enhance the collaboration and the interactions without secrets or borders among scientists all over the world. He has created organizations such as the World Laboratory and the World Federation of Scientists to make all this a reality. He has spent, and still spends, a large fraction of his energy to spread scientific culture, to make people understand that science and technology must be devoted to the benefit of mankind.

His actions have been based upon a deep love for science, knowledge and truth, and a deep faith in the Creator, the author of the logic of all phenomena in the universe, as Zichichi likes to say and write. This faith has always been manifest and has served as a motor for him, all along his life, to prove – with his concrete and constant endeavour – that «*science and faith*», as John Paul II used to say, «*are both gifts of God*». These words can be read, nicely engraved in an artistic plaque, at the "Ettore Majorana" Centre.

Robert A. Clark

The World Federation of Scientists (WFS), under its President, Professor Antonino Zichichi, in its basic mission to avoid planetary emergencies, has followed the goals and aspirations of His Holiness Pope John Paul II. The WFS has sought to apply science to improve human conditions, especially in the lesser developed regions of the world, through various innovative initiatives and programs. One of the very successful early demonstration projects, sponsored, was on the Yellow River in China.

The WFS Yellow River project involved development of a program for management and operation of the Yellow River. This project was undertaken in cooperation with the Yellow River Conservancy Commission (YRCC), the Academy of Sciences, and the Meteorological Service of the Peoples Republic of China (PRC).

The Yellow River, though much smaller than the adjacent Yangtze River to the south, is really the heartland of ancient China. Unfortunately, its resources were poorly managed beginning with the Qin and Han dynasties (221B.C.-220 A.D.), and by the 1990s the lower reaches of the river frequently experienced periods in which there was no flow and major irrigated areas were subject to severe drought conditions. In fact, in 1997 there was a period of 226 days with zero flow in the lower Yellow River and at its mouth.

The Yellow River Basin supports 12% of the present Chinese population of 1.3 billion people and has 15% of the cultivated fields nation-wide; however, the Yellow River (with average annual flow of 59 billion m^3) provides only 2% of the total national runoff. The Yellow River provides water for 50 large and medium-sized cities, and irrigated lands in the basin account for 36% of the cultivated area. The Yellow River facilities include 15 major reservoirs, 10,100 small reservoirs, numerous powerplants, and extensive irrigation infrastructure. In addition to improved hydraulic and hydrologic operations, hydropower generation in the basin has provided additional power to rural areas and cities in central China.

In the late 1980s, the Chinese agencies became aware of the WFS program under the direction of Professor Zichichi and initiated a request for support to improve operations of the Yellow River system. They asked WFS to assist them in the purchase of hydrometeorological equipment and development of a basin-wide forecasting system to manage the proposed Xiaolangdi Reservoir on the Yellow River above Zhengzhou. The program included support for the purchase of 25 weather stations that relayed meteorological data via satellite.

The programs supported by the WFS project also enabled the Chinese to obtain supplemental funding from other sources (e.g., World Bank) for construction of the Xiaolangdi Reservoir. That water project was placed in operation in 2001 and is functioning well. Operation of Xiaolandi has provided continuous flow to the lower Yellow River and increased irrigation and power production throughout the basin since 2001.

The WFS project also included training for many Chinese scientists and engineers in various scientific subjects such as hydrological and meteorological operations and forecasting, hydraulics, communications, data collection, remote sensing, and operation of a major river system for the development of irrigation, flood control, domestic and industrial water supply, and hydropower generation, in addition to integrating operations in this major river system as well as meteorological forecasting of heavy rainfall for hydrological prediction.

Training of personnel as a part of this WFS program provided the Yellow River Commission with scientists trained to operate the very complex Yellow River system. Many of these scientists are currently responsible for operation of the Yellow River facilities.

By fostering development of an operating system, Professor Zichichi provided substantial scientific support and technical assistance to the Chinese for the development and management of water resources of the Yellow River Basin, thus improving well-being for a large segment of the Chinese population.

The current Yellow River water facilities for the 795,000 km^2 basin is a demonstration of applied hydrologic science to a specific river basin that supports a population of more than 110 million people. Since completion of the WFS project, additional improvements to the lower Yellow River communications system have been provided by the Finnish and Chinese governments. Also, more data collection gages have been installed by YRCC above Xiaolangdi Reservoir.

After ten years of operation, most of the original hydrometeorological gages and weather stations are still functional, including the three original microwave networks. The WFS forecasting center, established in 1996, is now located in a new 25-story building housing YRCC headquarters in Zhengzhou The Yellow River program of the WFS demonstrates the importance of addressing critical, basic scientific components in the development of comprehensive national infrastructure and operation programs for conserving and managing water and other environmental resources in a responsible, sustainable manner.

The guidance and cooperation of Claude Manoli, Secretary General of the WFS, and his staff in the implementation of the Yellow River project are gratefully acknowledged. In addition, Gong Shiyang, formerly Director of the YRCC, and Niu Zhengqi, YRCC civil engineer responsible for implementation of the project, are also acknowledged.

Hassan Dalafi

The Majorana International Centre in Erice is considered to be one of the most important International Institutions in the world. This is due to the very many scientific and cultural events which have been organized in Erice during the last forty years.

Thousands upon thousands of scientists from all over the world have participated in the various activities of this Institute and obviously have benefited significantly. The activities have been mainly in the form of Seminars, Workshops, Schools and Conferences.

During the last few decades the Erice Centre has become a Mecca, in particular for the young scientists from Developing Countries and those from Eastern European Countries.

It is interesting to note, that from the beginning, the activities of the centre have covered the most important subjects in the cultural and scientific fields, always including fundamental arguments concerning world peace.

It is impossible to forget how much effort was made by Professor Antonino Zichichi and his close collaborators in Erice, Geneva and Lausanne in organizing those important, wonderful peace conferences, particulary during the period of the cold war, involving face to face discussions among the high level personalities from all over the world, in particular those from the United States of America and Russia. One could immagine how much courage was needed in organizing such meetings, taking into consideration the existing atmosphere during those years. The subjects of the meetings were, of course, focused mainly on nuclear debates and other relevant matters.

During the last few decades many conferences and seminars were also organized on the subjects concerning the Planetary Emergencies such as Climate, Water and Energy Emergencies. I remember with much pleasure those seminars which were organized every year during the period of 19th-24th of August with the participation of great scientists from various parts of the world, among them several Nobel Laureates.

I would also like to underline the importance of the World Laboratory and its programmes that provided many young scientists from the South and the East with fellowship and research grants, permitting them to follow their reserch activities in the most advanced institutions mainly in various European Countries.

It should also be mentioned that, within the activities of the World Laboratory, expert scientists could present projects in various fields to assist the institutions in developing and eastern countries. These projects were presented in the Council of the World Laboratory for approval. The Council had sixteen outstanding members, among them several Nobel Laureates. I remember with pleasure my participation in the Council Meetings, representing the Nobel Laureate Abdus Salam. It was truly an exciting event, with many discussions and battles for approving the projects presented.

The project leaders spent a period of time in those countries where their project was implemented, to personally follow the project. Furthermore, the World Laboratory very generously covered all the expenses involved, including the purchase of the technical and laboratory instruments forseen in the project.

I would like to finish this testimonial by emphasizing that all the above mentioned activities realized at the Erice Centre and the World Laboratory are undoubtedly, to a great extent, due to the strong leadership of their founder, namely Antonino Zichichi.

I would therefore suggest that on the occasion of his 80th birthday the Erice Centre be called Zichichi-Majorana International Centre, as it was the case of the Bohr Institute in Copenhagen.

Guy de Thé

It was in 1986 if I recall correctly, that Nino Zichichi contacted Bob Gallo, then at the National Cancer Institute, (NCI/NIH) in Bethesda, Md, USA, inviting him to conceive an international HIV/AIDS program under the aegis of the World Laboratory, which had been established a few years earlier to respond to the Planetary Emergencies. In turn Bob contacted me to ask if I was ready to collaborate with him in managing an AIDS research network in Africa, the epicenter at the origin of the AIDS epidemic.

Having extensive experience in carrying out multinational research in the field of viruses as causative agents in human cancers, in Africa, China and South East Asia, and having recently been involved in HIV-AIDS in the Ivory Coast, I was indeed ready to enter into a new World Lab venture, together with a number of preeminent colleagues from Europe and the USA.
But the key to my agreeing to engage in this venture, was my meeting with Antonino Zichichi. His personality, his never ending energy, his far reaching views, his enthusiasm, as well as his belief that Science is determinant in responding to the ever increasing world emergencies, was overwhelming. As we all know, Erice has been instrumental in avoiding a nuclear apocalypse by permitting nuclear physicists from the big four powers to meet and exchange views. This achievement alone merits the highest recognition! There are those who will remember the early days of the Erice conferences when the top scientists involved in the fierce race between the USA and Russia towards the A and H nuclear bombs, came together to get to know each other, to chat about science and society, and to have drinks together in the vaulted cave of the San Rocco monastery in Erice, mixed with Russian songs and music played by participants.

Nino loves challenges, whether it be the nuclear arms race of yesterday, or the HIV/AIDS epidemics of today.

Within our World Laboratory (WL) AIDS program, sessions were regularly organized during the yearly Planetary Emergencies conferences, together with a number of workshops, and we implemented a network on AIDS research and control in different countries in West and East Africa. We were the first to stress the feasibility of preventing the mother-to-child transmission of HIV by short term anti retroviral therapy of the pregnant mothers, thus permitting the mothers to breastfeed their babies and keep them free of HIV. In Nigeria the WL project led to an ongoing HIV vaccine preparedness trial in Jos, Central Nigeria, coordinated by the HIV institute of Bob Gallo in Baltimore.

Louis Pasteur used to say that "Knowledge is the treasure of mankind".

In the same vein, Antonino Zichichi is deeply convinced that Science is the best approach in facing crises and ensuring Humanity's progress. In which category of great achievers should we place Professor Zichichi? - Among the World's best renowned physicists; among the great science thinkers and writers; among the most enjoyable of theater players; among the very rare "honnète homme"? For me personally, Nino Zichichi is a sun, a cultural achievement of the sacred stones forming the multitude of churches in Erice that he transformed into temples of scientific knowledge, temples of the future.

Thank you Nino for being who you are. Thank you for the way you encourage us to listen to lectures far from our field of experience, and to think and act in a multidisciplinary way. Thank you for your unique way to be a man we greatly admire and are so very fond of!

I do indeed consider it as a unique "privilege" to be a friend of yours.

Carmen Difiglio

Professor Antonino Zichichi has established a unique interdisciplinary forum in which the world's most pressing problems are discussed. The aim is to find scientific solutions and to encourage action. Since these problems can not be solved by a single scientific discipline (nor can they be solved without a view from the social and political sciences), the Seminars' participants constitute a broad scope of expertise from the physical sciences, the social sciences and policy makers. The interaction among such diverse experts ensure that problems are not oversimplified by viewing them from a single perspective.

While my participation has been relatively recent (these seminars have been conducted for almost four decades), my understanding is that their initial focus was on the most pressing planetary emergency of the time: avoiding nuclear war. While potential nuclear conflicts are still discussed, since the end of the Cold War, many other planetary emergencies have become a staple of the Erice Seminars, including the energy and climate change, topics that have drawn me to the Seminars.

A key resource of the Seminars are the several Permanent Monitoring Panels that bring the experts together to define the problems and possible solutions. Their work feeds the discussion by all the participants at the Plenary. The debate involves scientists from all parts of the world. As low income, poor health and lack of self-development opportunities are inequitably distributed, it would be meaningless to discuss these issues without participation by experts from the full range of countries: North and South, East and West.

I have had the opportunity to promote interdisciplinary study of energy. It is interesting how energy is a factor in so many of the world's problems: with energy you can provide healthy water supplies to communities that do not have them; with clean energy you can reduce pollution in all parts of the world, with energy you can provide light in communities that go dark after the sun sets and power to raise their standard of living; with energy you can maintain the world's economic growth and prosperity. The Erice Seminars provide a unique opportunity to focus on these opportunities.

None of this would work without an inspiring leader. Professor Zichichi has devoted his life to making science an instrument for solving the world's problems instead of contributing to them. I am old enough to remember how the development of nuclear weapons placed the world at a crossroads, whether humanity itself would survive. Scientists were at the center of the crisis. Professor Zichichi engaged scientists from Lawrence Livermore and their Soviet counterparts to discuss ways to verify arms control treaties. President Reagan's "trust and verify" illustrates how important these discussions may have been to achieving the INF treaty and an end of the nuclear arms race. While the current planetary emergencies may not be as dramatic, Professor Zichichi's leadership still permeates the spirit of Erice.

Mbareck Diop

I met Professor Antonino Zichichi twelve years ago during a World Laboratory meeting in Geneva, attended by, amongst others, the eminent Scientists and Nobel Prize laureates, Kai Siegbahn and T.D. Lee.

I have also attended the Erice meetings on Planetary Emergencies and have been impressed by such simplicity and scientific culture. I was immediately inspired by the ideas of "Science without frontiers", "Science for peace" and "Science for development and for a healthy world".

Coming from Africa, the poorest continent, I was encouraged by the Erice spirit. After each meeting of the Planetary Emergency Panels (PMPs), I could not imagine how our world could overcome challenges such as the food, the energy, the environmental crises, and other Planetary Emergencies as defined by Professor Zichichi, without the valuable work of major scientists from all continents.

Involving the scientific community without any distinction in race, religion and ideology, in the research of solutions to improve the world, is remarkable, and Nino has been a superb leader in this accomplishment.

Nino's vision is underlined by interdisciplinarity. Yes! He is absolutely right. I was smiling when Nino repeated with conviction, this magic word. We cannot make progress in Science without interdisciplinarity.

My other great admiration of the World Laboratory and the World Federation of Scientists is the continuous dialogue between East and West Scientists in order to avoid confrontation and a nuclear war.

One cannot measure the huge contribution of Erice dialogue in preventing the world from holocaust.

Putting together major Scientists from each block and engaging in sincere discussion on hot topics in the spirit of Erice has helped to decrease tension between the super power states. Nino is admirable in his ability to conduct exchanges in a scientific spirit, without violence.

I cannot end my short comment without mentioning the considerable contribution of the National Scholarship Programme of the World Federation of Scientists in the fight against the brain drain which is negatively affecting the developing world. Young Scientists awarded the National Scholarship Programme are in a position to fully conduct their research in their home country, supervised by local Professors, and this with the financial support of the World Federation of Scientists National Scholarship Programme.

Lorne G. Everett

For two decades it has been my distinct pleasure and honor to participate in the International Seminars on Nuclear War and Planetary Emergencies in Erice, Italy. While the issue of nuclear waste stewardship and repository design has been fundamental to the pollution panel, a number of pollution emergencies have been thoroughly vetted over the years. Beginning with the disposal of municipal wastes, petroleum hydrocarbons, chlorinated hydrocarbons, plastics, and now pharmaceutical wastes, the pollution panel has sought an understanding and developed a series of recommendations relative to the above planetary emergencies. This back end-of-the-pipe approach, however, has now lent itself to a 2009 Green Chemistry approach which is a front end-of-the-pipe realization that getting hazardous materials out of the system prior to its disposal should be the path forward. Underlying and interspersed with many of the pollution panel meetings were issues related to high-level radioactive waste repositories and the commensurate issues associated with mining, milling, delivery, risk, and security. Early in our meetings, I conceived of a Memo of Understanding (MOU) amongst representatives of several nuclear powers including Japan, United Kingdom, Russia, Canada and the United States on a Long-Term High-Level Radioactive Waste Stewardship Program. That draft MOU has contributed to the current Panel recommendations for the development of a multinational high-level radioactive waste repository which is proposed to be managed and safeguarded by all members of the nuclear energy community.

Highlights of my Erice experience over the years include the distinct honor of spending one week in "la sala de Pope," his Holiness Pope John Paul's apartment in the San Francesco Monastery in Erice. Sleeping in the Pope's bed and looking up at the magnificent cupola in the main room was truly a spiritual encounter.

With great humility I recall receiving the Piotr Kapitza Gold Medal at the Beau Rivage Palace in Lausanne Switzerland in 1999 from Dr. Igor S. Zektser Head of the Water Problems Institute in the Russian Academy of Sciences. Piotr Kapitza, a Nobel Laureate, was one of the three original signatories to the Erice Statement. It is underappreciated and mostly unknown throughout the western world that the Russian Academy has chosen to name its highest award for original research after a nuclear scientist who refused to build the atomic bomb.

No one will forget when Dr. Edward Teller joined the meeting hall in Erice by video link and he shared his giant intellect by asking the audience for any question on the energy theme. Of particular note in that video broadcast, Dr. Teller recommended that the world should follow the use of developing nuclear energy by locating nuclear power plants 200 feet below the land surface and upon the completion of their useful life, the cavity, including all of the infrastructure and radioactive waste should be buried in place. When Dr. Teller was no longer able to participate in the meetings in Erice, he would send his envoy, Dr. Dixie Lee Ray, former Governor of the State of Washington, USA and the last head of the Atomic Energy Commission. I recall sitting next to Dixie the morning after we had enjoyed ample quantities of wine in the Marsala cellar. Several members of the European community were commenting in the General Assembly on the extensive waste that resided at the Hanford Site in Washington State as a result of nuclear weapons development. Dixie stood up sharply, and commented that she needed to "speak on behalf of all of those folks who had gone before her." She said that she had several months to build a nuclear bomb. And in so doing had created some residual contamination, however, she continued forcefully, had she not built the bomb “you all would be speaking German.” Dixie sat down and the room was as quite as I had ever heard it.

One noteworthy heated exchange took place at CERN when Nobel Laureate, Kai Siegbaum, as session moderator insisted that Dr Robert Gallo (co-discoverer of the HIV virus) should leave the stage since his speaking time was up. Dr Gallo emotionally retorted that he wanted to say more about the pandemic of AIDS in Africa and that saving time for physicists to talk about subatomic particles that do not exist was not the best use of the time.

Several times I was staggered at the level of conversation at the various dinners held in Erice. On one occasion Gregory Canavan, in a soft voice, simply said that he, Edward Teller, and Lowell Wood had won the cold war. Since Edward Teller was the Science Advisor to Ronald Regan, and the main proponent of the Strategic Defense Initiative (Star Wars), this statement of monumental importance was presented in a matter of fact discussion. On another occasion David Scott asked if he could join my dinner table. After a couple of glasses of wine, he indicated that he was Commander David Scott who had Commanded Apollo Flight 15 (and the seventh man to walk on the moon). David indicated that he had accumulated great wealth by designing a space tug concept to use rockets that could attach to wayward satellites to bring them back into orbit, an approach that he sold for a small fortune to international insurance companies. David indicated that he wanted to take that financial depth and apply it towards encapsulating high-level radioactive waste and sending it on rockets to the sun to be destroyed by fusion. I indicated that I personally, and the audience in general, may not be supportive of such an approach yet we were willing to listen to his highly creative ideas.

On every occasion in my 19 years of continuous travel to Italy, I was staggered at the breadth and depth of the intellectual understanding of Dr. Nino Zichichi. My feelings for Nino are only surpassed by the warmth, grace and charm of his wife Maria.

No testimony about Erice would be complete without recognizing the coordinating skills, management insights, and technical awareness provided by Claude Manoli and his staff. Claude’s devotion to his President has been total, complete, and a major underpinning of the enormous international success of the Erice meetings.

Paolo Fulci

I met Antonino Zichichi for the first time in 1976. Senator Amintori Fanfani had been re-elected president of the Senate and recalled me from our Tokyo embassy to put me in charge of his Secretariat.

I had just started my duties at the splendid Palazzo Madama when, one morning, Fanfani asked me to his office. With that sly smile of his that he wore when he was in a good mood he asked me: "I assume you have already met the most famous son of your Sicilian Island?". I did some rapid thinking (with Fanfani answers had to be immediate) and the first two names which came to mind were Vittorio Emmanuele Orlando and Gaetano Martino.

"No – replied Fanfani – those two don't exist any more. I refer to the living, and in particular to the famous Antonino Zichichi". I had to admit that I had not had the privilege of knowing, outside of his fame, this illustrious son of Sicily. This situation did not last, as the very same night, at two in the morning, I was awakened by an excited voice saying "Zichichi speaking. President Fanfani has referred me to you. I have to discuss with you an urgent matter".

Briefly put. While I was trying to emerge from a deep sleep, he explained to me that there was an urgent need to contact the president of the Abruzzo region. The tunnel of the Gran Sasso of the A24 Highways was being drilled, and it gave Zichichi the idea of creating, next to the tunnel, a large underground laboratory for the study of elementary particles. Taking advantage of the drilling, a vast area could be set out, under 1'400 meters of rock, for what would become – and still is today – the deepest underground laboratory in the world, the Gran Sasso National Laboratory. Zichichi's worry was that the drilling would proceed beyond the strategic location of the laboratory, which would complicate enormously the realisation of the Laboratory diggings. I promised Zichichi that I would call the regional president first thing in the morning but he kept insisting that I place the call immediately! I tried my best to persuade him that waking up the president, in the middle of the night, even for such an important motive would seem bizarre and could prove to be counterproductive.

But Zichichi continued insisting! In my whole life, I've rarely met a person who could beat Zichichi in the application of Disraeli's saying: "In life, the secret of success lies in the constancy of the proposals". I called the president the following morning who immediately understood the importance of the proposal and started its realisation. If an advanced laboratory such as the Gran Sasso exists today, it is largely due to Nino Zichichi.

In 1980, I was designated ambassador to Canada. My main task in this North American "Island of Happiness" was to promote the image of Italy in all domains. I immediately thought of Zichichi and his talents as a communicator. Despite the cold of mid-January 1981, he immediately accepted to come and give public conferences in Montreal, Toronto and Ottawa, on the frontiers of modern physics, the relation between Science and Faith and on Galilei's scientific methodology. One of Zichichi's many talents is to make the most abstruse scientific theme easy to understand and even enthralling. The public was bewitched, especially the Italo-Canadians, so proud of him that they gave him, in Toronto particularly, a triumphal welcome.

In summer of that same year, Nino invited my wife and I to Erice. We were delighted by the beauty of Erice and impressed by Nino's success in giving back a soul – a sort of *Renaissance*

- to that medieval town, including in its churches, convents and monasteries which had been turned into pulsating centres of the most advanced scientific thinking. We were also very impressed by his capacity to attract an incredible number of scientists, many of them Nobel Prize winners, who kept meeting day after day, year after year, to investigate themes of the utmost importance to humanity, such as the nuclear war scourge and its apocalyptic consequences, by the very high tenor and fervour of their discussions and the constant interweaving of new links and friendships in the scientific community, without distinction between race, language, religion or political appartenance.

I was so interested by what was going on in Erice that, in August 1983, I suggested to the Canadian Prime Minister Pierre Trudeau, whom I was accompanying in his brief Sicilian vacation, to come and visit the Ettore Majorana Centre in Erice.

The visit took place on 21 August 1983 and the theme for the Seminar was "Technical bases for Peace". On that occasion, Trudeau told me "What they're doing here in Erice is surely the most relevant and factual contribution provided by scientists, from the world over, to the development of Peace". This happened in the period during which the most extreme measures were being taken in the crazy armament race between the West and the Soviet Union, each vying for being the most powerfully armed. The race reached its summit in 1985 when, following the deployment of the first Euro missiles to counter those of the Soviet Union, the crisis between the two blocks reached its peak.

Trudeau spoke in Erice, reminding that Canada, despite having large Uranium deposits, always refused to produce atomic weapons. Canada was firmly against the development of any new weapons, even those labelled as being "defensive", as it would threaten the strategic statu quo and divert the existing resources from their pacific use for the benefit of humanity..

Back in Erice, Trudeau told his collaborators and ambassadors of many nations how favourably impressed he had been by what he had seen and heard in Erice: a small UN for Science!

In the years that followed, I kept in contact with Nino Zichichi but not as much as I would have wished. It was Zichichi again, in the late eighties, when I was the permanent Italian Representative in NATO, who convinced me to strongly and successfully oppose, with the approval the President of the Council, the Hon. Giulio Andreotti, the request of certain members to drastically curtail the funds allocated to scientific research, the so-called NATO "third dimension".

Finally, as a perfect illustration of Zichichi's personality, I'll mention an article which appeared in July 1985 in the "New Scientists". Its title could not have been more fitting or significant: "How the Italians put whizz into atoms: Antonino Zichichi, a scientific champion of the kind that Britain desperately needs".

William Fulkerson

It is an honor to be invited to participate in the International Seminars on Planetary Emergencies at Erice sponsored by the World Federation of Scientists (WFS).
I am old but not an old timer at Erice. I have participated only since 02, but it has been a life changing experience, primarily because of the people. I was invited through the good offices of Bruce Stram who was reconstructing the Energy PMP in 02. There I also met Dick Wilson who was elected chair of the "new" Energy PMP. Dick has a balanced, wise, and insightful appreciation of the global energy problems we face which are getting worse, and I learned from him.

With 9/11 on my mind, my first plenary session talk at Erice was titled "Oil Policies and Terror: Getting Off Oil." Thus, pushed into the terrorism crowd as well as the energy crowd, I met Dick Garwin, brilliant on non-proliferation and everything technical about Weapons of Mass Destruction, Scott Atran with the courage to talk to terrorists to learn their motives, and Ahmad Kamal, former Ambassador to the UN from Pakistan, who brought a developing world point of view. Despite the fact that poverty does not seem to be a prominent cause of terror, powerlessness is. Terror becomes a tool for the weak against the powerful in an asymmetric struggle. With Lord Alderdice, of Northern Ireland fame, chairing the effort, the Erice studies on terrorism will be important. He taught us that "you don't dis (disrespect) your enemies" if you want to find peace.

It is the integration of energy with so many other planetary emergency themes that makes the Erice Seminars so important to me. Gradually the attention of the Seminars has changed from East/West confrontation to the North/South dilemma. The Energy PMP's concern about the role of energy in the process of development is intense despite the fact that eliminating energy poverty was not made one of the U.N. Millennium Development Goals. It should have been. Several of us began to consider the problem of universal electrification that must be done sustainably in the sense of allowing little or no carbon emissions. The 1.5 billion people who live without electricity are at a huge disadvantage. The rapid growth of many developing nations including India and China is to be applauded, but it does and will continue to drive anthropogenic climate change. This is the clear predicament facing the world, balancing economic growth against the changes to the energy system necessary to prevent dangerous interference with the climate system.

Recognizing this predicament, the Energy PMP collaborated with the Climate PMP and the Limits for Development PMP to organize a plenary session and workshop in 07 on a Framework for Managing Climate Change. The Chairs were Art Rosenfeld, Energy, Bill Sprigg, Climate, and Geraldo Serra, Limits for Development. This collaboration would have been impossible without the help and good council of Claude Manoli, Secretary General of the WFS. Several important recommendations for the world community of nations emerged from the workshop including one that research on geoengineering should be adequately funded. The Framework also became the basis for a technical session at the Vatican on the occasion of the awarding of the Erice Prizes for Peace in December of 07.

Erice is a beautiful, historic place, and we, who are lucky enough to be participants, are pampered and made to feel special by the WFS staff, by the people of the Ettore Majorana Foundation and Centre for Scientific Culture and by the people of the town itself. The staff, with their attention to detail and their kindness, make the experience enjoyable as well as stimulating.

Of course, the person who most embodies the Seminars is Professor Antonino Zichichi himself, the President of WFS. He sits on the first row aisle seat during the Plenary Sessions, usually right next to Professor and Nobel Laureate T.D. Lee, his co-chair, and he keeps things energized and exciting with comments and questions. In the evenings he presides at each banquet as a sort of benign lord keeping the wine and talk flowing. He has made the Seminars a truly unique phenomenon in the arena of international exchange of ideas and knowledge about Planetary Emergencies.

Bertil Galland

A phone call from Antonino Zichichi in 1983 brought me his first invitation to Erice. Both of us lived in Switzerland. We had never met. He had his home in Lausanne, where I worked as a journalist for *24 heures.* In this newspaper I reported about wars in Israel or in Vietnam, presidential primaries in America or the last throes of the cultural revolution in China. I knew Zichichi as one of the eminent scientists at the CERN in Geneva. He told me he had noticed my name among the recipients of an honorary doctorate given that year by the University of Zurich. The award was related to publishing activities in Switzerland and had nothing to do with physics or world affairs.

With a background in political science, it was never clear to me why I should have had the privilege to participate, every year since 1983, in the August seminars at the Ettore Majorana Center in Erice, among some of the world's most renowned nuclear physicists. Zichichi's motivation was perhaps simply to have me as a witness of those historical encounters. My only contribution has been to sit, watch and listen. Then I wrote about it in the Swiss press. From the beginning, at the third Session of the "International seminar on nuclear war", I understood that I had been led into a hot spot where events of great magnitude were taking shape. Among the participants was Edward Teller, father of the hydrogen bomb, 75 years old. He was one of the main scientific advisers to President Reagan. I could see his imposing figure, cane in hand, facing the fiery and much younger Evghenii Velikhov, 48, head of the Soviet Academy of sciences and principal actor in the Kremlin's military and civil nuclear program.

The cold war, in 1983, rather than settling towards the pacified grounds opened by the bilateral disarmament negotiations, was suddenly rebounding. In March, the President of the United Sates had shaken the world with the announcement of his Strategic Defense Initiative. The leading thinker and actor of the explosive program, referred to as Star Wars, was Teller. The Soviet authorities and Velikhov, their top scientist, were in a state of shock. The balance of terror was on the verge of tilting in favour of the USA. In Erice's cobblestoned streets, we could feel, behind the unusual number of Italian police in uniform, a planetary tension.

But the lectures and discussions on "the technical basis for peace" proceeded according to the schedule set up with graceful authority by Zichichi in San Domenico, the aula magna, and during the evening meals where East and West easily found common pleasure in Sicilian foods and velvety wines. In accordance with the fashion of the early eighties, the lord of Erice wore his hair long and wavy. Participants of all creeds and nations, but with impeccable scientific credentials in the most varied fields, were admitted to abstruse and technical exchanges centering on the risk of our planet being destroyed by a nuclear confrontation. But behind the intricate experimental endeavours and wide cosmic considerations, one could feel the warmth of a club.

The inner circle in Erice was a very select group of nuclear physicists with close government links. We enjoyed the presence and the contributions of a good number of Nobel Prize winners. We shared our days, for instance, with another Hungarian-born American and star of nuclear physics, Eugene Wigner, 81 years old, or with T. D. Lee, 57, who had been one of Enrico Fermi's favorite students and brought back to Italy an echo of his genius. There was no exclusion. Each participant could share with all others the formal or open discussions in San Domenico, San Francesco or San Rocco, the three former monasteries of the Ettore Majorana

Center for scientific culture. With stars of this brightness and in this rare intellectual intimacy, even coffee breaks with fruit, panini and a view on Capo San Vito and the Tyrrhenian Sea took on an extraordinary dimension.

Beneath the technicalities of the Erice lectures a fact of contemporary history became evident: physicists, after World War II, had become, in a way, the masters of the planet. Familiar with tremendous new forces, not fully grasped by others, they could advise and strongly influence the superpowers, thereby multiplying the impact of their military forces. By the same token they had their reasons to warn the World leaders against catastrophic misuse of the new weapons. In the eighties, no other category of counselor had maintained such weight in Washington and Moscow. With the scientific and human ties Zichichi had woven among his fellow physicists around the world, especially through his contacts in Britain with Paul Dirac, it was possible to address the question of why Bertrand Russell, as a great thinker, had been so miffed until his death. Russell had tried himself to play a central role in planetary affairs and wondered how the physicists had usurped, as main advisers to the superpowers, a position that, in his view, should only belong to the philosophers.

However, Piotr Kapitsa, the Russian Nobel prize winner in physics in 1978, had actually refused to devote his nuclear know-how to Soviet nuclear strategy and its tremendously risky developments. He was absent from Erice in 1983 and close to his death in Moscow. But the preceding year, with Dirac and Zichichi, he had signed, with exceptional courage, the famous Erice Statement. The trio had warned the World of a nuclear holocaust and appealed to science for peace, with no borders or secrets. Strong language at the time of the big game between the CIA and the KGB.

Teller was among us in Erice, one hand holding his cane and the other on his disciple Lowell Wood's shoulder. The young redhead professor was an intense person with scientific excellence, cutting opinions and a passion for bold technical inventions. The two physicists were working in America by order of President Reagan, who felt that his country was not properly protected against Soviet strategic nuclear ballistic missiles. "And he was right," Teller said to me haughtily one evening when I happened to be sitting beside him during supper at the Ristorante Elimo. I had asked him whether the first idea of the SDI came from him or from the President. The father of the hydrogen bomb conceded the paternity of the new project to Reagan, but we read in the press that he and Peter Hagelstein were bringing forth, at the Lawrence Livermore National Laboratory, the concept of an array of satellites capable of detecting and destroying, with explosion-powered X-laser, any hostile intercontinental missile.

Among the American nuclear scientists and astrophysicists, quite a number expressed doubts about the system's feasibility and opportunity. Richard Garwin, 55 years old, one of the brightest and the friendliest among them, was in Erice. I remember a conversation with him on the terrace of the Ristorante Ulisse, where he very kindly rephrased the reasons for his reservations at my modest scientific level. I heard later that in the US he paid the price of his scientific cautiousness.

But the main attack against the SDI was brought, of course, by Evgheni Velikhov. In one of Erice's most tense moments, the speech he delivered in the packed aula magna was vehement. He gave, from what I recall (but I was unable to follow his mathematical demonstrations on the screen), three reasons to condemn Teller's great leap forward to suit President Reagan's wishes. Velikhov's half bald forehead was sweating while he expressed his physical and scientific strength with concentrated and forceful gestures. First, I remember his saying that the SDI was a breach of international agreement. This initiative was challenging the trust painfully cemented between the US and the Soviet Union. It was a reckless threat to the balance between the two superpowers. And it went against the spirit of Erice. Secondly, Velikhov said, Teller's system could not work. And he started to prove it by scribbling cryptic figures and formulas. Awestruck, I watched them trickle down on the screen above him.

I was surprised to hear that a non-functional system could be considered a catastrophic threat. But during his discourse, Velikhov talked to his fellow physicists with reasoning I could not grasp. I became even more confused by the Soviet academician's third argument. "Nothing in the SDI is a surprise for us," he said in substance. "Our scientists have been actively working in this field, and technically we are perfectly able to produce the same system, and in fact we are already on the verge of implementing it."

Well, well, I thought. Scientifically then, Star Wars, according to the Russians, is 1) reprehensible, 2) impossible, 3) and already being developed in the Soviet laboratories. According to my humble logic something seemed wrong. Velikhov, in this accumulation of contradictory arguments, had, above all, given a strong signal of the Kremlin's disarray.

Everybody was waiting for Teller's answer. It turned into the shortest and the most amazing speech I have ever heard in Erice. Slowly the American physicist rose and moved his towering and aging body from the first row of the listeners to the steps leading to the speaker's stand. In the right hand he held his long wooded cane. It helped him walk, as one of his feet had been harmed by a streetcar accident in Germany during his pre-nazi studies. He stood up with the microphone hanging around his neck and dangling against his ruffled tie. He seemed to defy the assembly. His eyes, slightly veiled under his bushy eyebrows, were directed at his Russian friends in the front rows. They knew that the man facing them was a formidable American scientist, but also a dear colleague of Hungarian origin, fully aware of all the tormented history of his native country and one of the most convinced haters of the communist regime.

His address was a short story. "The other day," Teller started in his slow, deep, theatrical voice, putting his weight on each word, "I was in Montana. I had been invited to give a lecture. I thought it was scheduled for the morning. But it was programmed for the afternoon, which meant I was free. Montana is a state covered with big forests, so I decided to go for a walk in the woods. My friends warned me: 'Don't go, there are grizzlies out there !' I answered: 'I've got my stick.' They insisted: 'The grizzly bear is an enormous animal.' I showed my cane. 'Please believe us,' they said, 'your cane cannot protect you against the beast.' I answered (Teller lifted his cane which appeared beside his body on the wall casting an impressive shadow): 'I know it. But the grizzly bear doesn't."

That was Teller's answer to the Russians in Erice. The father of the hydrogen bomb took off his microphone, left the stand and stepped down to his seat.

Later that day, not knowing what a journalist could reasonably write after a scientific exchange of this kind, I seized the opportunity of a short confidential talk with Zichichi and begged him to tell me just one thing. In his opinion, could the SDI system work or not? He laughed at my naive question and said: "If the United States suddenly devotes billions of dollars to develop three fields – laser techniques, optical hyperprecision and miniaturization of computing to put it on small satellites – that is enough to quickly bring them so far ahead that they will be unreachable by the Soviet Union. The question is not whether the system will ever work."

The rapid meltdown of the Soviet Empire in the following years confirmed that Zichichi was right. I never forgot the grizzly bear and Teller's story. It was one of the most brilliant lessons I received in Erice, not in physics but in political science and in contemporary history. The push of a serious threat, founded on unexpected research and development, is sufficient, without experimental proof of its workability, to throw a big power off balance.

Richard L. Garwin

I first met Antonino ("Nino") Zichichi in 1958 when I was in Geneva for six weeks on assignment to the U.S. government delegation to the U.N.-sponsored "Conference on the Prevention of Surprise Attack" and participated also in the Conference of Experts on Detection of a Nuclear Explosive Testing. Leon Lederman, then at Columbia University, was spending a year at CERN as a Ford Foundation Fellow and had organized an exploratory team to pursue the precision measurement of g-2 for the muon. I had occasion to bootleg time on the CERN computer (a Ferranti Mercury) and no doubt used the occasion to look up Leon and to see what he was doing.

Although I had worked in particle physics after receiving my Ph.D. in 1949, I left that field in December 1952 when I moved to the IBM Laboratory at Columbia University and changed my field to condensed matter physics-- liquid and solid helium and superconductors. But I had developed several of the important techniques in high energy physics and had even brought with me to New York a coincidence-anticoincidence analyzer that I had the electronics shop at the University of Chicago make, simply because I thought that this technology transfer to Columbia could be useful.

When in the summer of 1956, T.D Lee and C.N. Yang wrote their theoretical paper on the possibility of parity violation in the weak interactions, they worked out the consequences for the pi-mu-e decay chain, as well as for the decay of Co-60. Ultimately, Leon and I and Marcel Weinrich did an experiment in four days elapsed time in early January 1957, established massive parity violation in the pi-mu-e decay chain. We used the electronics I had brought from Chicago.

Our first experiment determined the precession rate of the muon spin in a magnetic field, but it was also a crude determination that "g-2" ~ 0, because the muon spin direction with respect to its velocity was unaffected by its passage through the fringing field of the cyclotron, thus setting the spin of the muon as ½.

Although the next couple of years were filled with our efforts to determine more and more accurately the magnetic moment of the muon, the quantity of greatest theoretical interest was g-2, for which a very accurate value of the muon mass was required, if it was to be computed as the ratio of the measured magnetic moment to the Dirac magnetic moment for a particle of muon mass.

It was clear that we could perform a direct measurement of g-2 if we could store muons in orbits for many turns in a magnetic field, or, equivalently, for a long time at high field. The muon lifetime (2.2 microseconds for a muon at rest) put a limit on the accuracy that could be obtained, but when Leon went to CERN in the Fall of 1958, he organized a group with enthusiasm to conduct such an experiment using the proton synchrocyclotron that was just being put into service.

CERN seemed like such an attractive place that I also sought support from the Ford Foundation to spend the year 1959-60 there, but I just wanted to read in the library to refresh my knowledge of physics, somewhat depleted by work that I had been doing in Washington for several years, primarily on military technology.

It was not to be, because Leon's nascent g-2 group keenly felt the absence of an experienced leader such as Leon Lederman, and besieged me with pleas that I should lead the experiment. Ultimately, Gilberto Bernardini, whose office was across the hall from mine, took up the siege and in his soulful way obtained my agreement.

The group consisted of Georges Charpak, Francis J.M. Farley, Richard L. Garwin, T. Muller, J.C. Sens, and Nino Zichichi. Initially Carl York and Val Telegdi were involved, as well. Since I had been asked to lead the effort, after considerable discussion, it was agreed that individuals in the group would have specific responsibilities. Charpak's was the counters that would determine the polarization of muons emerging from storage in a large magnet. Zichichi's responsibility was to accurately shim the magnetic field of this magnet—ultimately six meters long and 80 tons in mass—so as to produce the orbit paths necessary for an injection region, a storage region, and an extraction region. The magnet still existed at CERN as of 2005, as described in a touching article by Chris Jones.[1] Jones emphasized the "clever work" associated with the shimming to provide the orbit advance of 2 cm/turn for capture upon entry to the magnet, then 0.4 cm/turn for maximum storage time and ultimately 11 cm/turn for the ejection region.

Although there was daily interaction among the group, we had a formal meeting each week to report progress, and the experiment worked extremely well.

Francis Farley went on to design experiments of much higher precision at CERN and at Brookhaven, taking advantage not only of the relativistic life extension of muons but also of various "magic energies" and other aspects to improve our measurement of (g-2)/2 from 1162(5) x 10^{-6} to 1165.9208(6) x 10^{-6}.

Nino was obviously a highly capable and energetic physicist, and I was delighted when he soon asked me to be on the Board of a new organization that he was founding in Erice, initially to host NATO summer schools. In fact, I organized and led one such school for a week in 1966, as I recall, on "Microelectronics," while I was Director of Applied Research at the IBM Research Center, Yorktown Heights.

I was very busy with my IBM technology, my research in physics, and my work for the U.S. government, so I did not participate much in the Erice activities, which, perhaps for the reason, grew very well. In 1981-83, though, I spent a week each summer in vigorous participation in seminars in Erice, "International Seminars on Nuclear War."

These brought together from the United States Edward Teller, Eugene Wigner, W.K.H. Panofsky, Spurgeon Keeny, and myself and others, and from France Georges Charpak, and from the Soviet Union Evgenii P. Velikhov and others.

Discussion was often well informed, informal, and at times passionate.

I had worked with Edward Teller on the first test of a hydrogen bomb, MIKE, November 1, 1952, which exploded with a yield of about 11 megatons, almost 1000 times the energy produced by the bomb dropped on Hiroshima. In fact, at the seminar in 1981 Teller stated

---

[1] http://cerncourier.com/cws/article/cern/29489

"I want to do so by telling you a story of which I believe no one has heard. In the early 1950's when I had the first crude design of the hydrogen bomb, Dick Garwin came to Los Alamos and asked me how he could help. Actually the design I had in mind was not that of a real bomb but of a model for an experiment. I asked Garwin to change this crude design into something approximating a blueprint. He did so in a short time-- a week or two. That experiment was carried out. Garwin's blueprint had been criticized by many people, including Hans Bethe. In the end the shot was fired almost precisely according to Garwin's design, and it worked as expected."[2]

Spurgeon Keeny, on the American side, had been a key person on the National Security Council staff and with the President's Science Advisory Committee beginning in the mid 1950s. Eventually he was to become Deputy Director of the United States Arms Control and Disarmament Agency, and in fact was to be the administrative host of a small committee led by Paul Doty of Harvard University that included also Sidney D. Drell, W.K.H. Panofsky, myself, and Jack P. Ruina and George Rathjens of MIT. This group would meet monthly with Henry A. Kissinger as President's Nixon's National Security Advisor or later in his role as both National Security Advisor and Secretary of State.

Keeny was intimately familiar with the U.S. and Soviet offensive and defensive technology and also with the details of the arms control negotiations between the United States and the Soviet Union to try to stabilize their threat.

Edward Teller, of course, had worked at Los Alamos on the hydrogen bomb and then in 1951 moved to the nascent nuclear weapons laboratory at Livermore, California. He and Eugene Wigner were active in encouraging the development and deployment not only of active defenses against Soviet nuclear bombs and missiles, but also passive ("civil") defenses such as blast and fallout shelters. Panofsky, Keeny, and I opposed the deployment of any active defense that had thus far been considered, because they could all predictably and continually be defeated by relatively simple countermeasures or tactics on the part of the Soviet Union.

Harold Agnew had just finished his 10-year term (1970-79) as Director of Los Alamos National Laboratory, and had had a long career there beginning when the Laboratory was opened in 1943. Harold Agnew had been the Project Manager for the first stockpiled U.S. thermonuclear weapon.

There was considerable acrimony in discussions on the American side, but much real information was exchanged, and the non-American participants were able to make their own judgments from this discussion. For instance, in the 1983 Erice volume (p. 16) Edward Teller comments "... in my opinion, ... to establish satellites in space is not an effective way of defense. In fact on March 23 (1983) when the President spoke of defense, the word 'space' did not occur even once ...". But there is also a brief presentation in which I demonstrated that Edward's alternative to space—a pop-up interceptor equipped with nuclear-pumped x-ray lasers—could be trivially defeated by a Soviet fast-burn booster, and that a fast-burn interceptor would be of absolutely no help in recovering capability. The counter to a fast-burn booster is not a more responsive interceptor but an interceptor with higher speed. Cutting in half the burn time of an ICBM can be done at a cost of about 5% of payload,

---

[2] Anecdote about R.L. Garwin's role in developing the hydrogen bomb. stated by E. Teller at the International Seminar on the World-Wide Implications of a Nuclear War, E. Majorana Centre for Scientific Culture, ERICE, August 14-19, 1981

whereas doubling the speed of the interceptor (which is what would be required) may increase the required launch mass from 100 times that of the payload to 10,000 times.

As an example of the breathtaking substance of some of the discussions one has former Prime Minister Fanfani's contribution in 1981.[3] "Negotiation needs means: an example." Here Fanfani reveals something that was denied by U.S. officials to the present day:

> "President John F. Kennedy asked me, as the then Prime Minister of Italy, for the withdrawal of the missiles in Italy. I asked the Turkish government if they would agree to the withdrawal of the missiles in Turkey. The position of Italy was that for the cause of world peace at that moment it was better to withdraw the missiles rather than fire them. In any case, Italy would consent to the withdrawal only if Turkey did. These two countries together agreed to withdrawal of the missiles installed in their territories."

Despite the presence of journalists, Keeny emphasizes that no word of this revelation reached the Press or even since the publication of the Proceedings in 1982 has been known to any of the interested parties.

Edward Teller was a frequent participant in Erice, becoming quite close to his Soviet counterparts, especially Evgenii Velikhov. On 23 August 1983, the two of them, with Zichichi, signed a document "to establish (at Erice) a joint USA-USSR and European research group to study the simulation and evaluations of global consequences of a USA-USSR nuclear confrontation, and the way out of the present balance of terror and … if it is possible to conceive a new type of defense system against nuclear destruction."

Edward was well known for his vocal advocacy of abolition of secrecy, and also for his use of secrecy to deny cogent facts to those who might oppose his programs. In 1983 Edward was accompanied to the meetings by former Atomic Energy Commissioner Dixy Lee Ray, who then became governor of the state of Washington. Dr. Ray called me a "traitor" for my willingness to present at Erice my arguments against the Strategic Defense Initiative of the United States, which was fervently supported by Edward Teller, because I was speaking abroad in opposition to the policy of the United States. My participation in the meeting would have been hollow if I had been unwilling to present in Erice what I had already done in my open writings and testimony in the United States in attempting to create a sensible policy.

This time, two columnists for the New York Times and the Washington Post (Flora Lewis and Mary McGrory) were there, and both covered the incident in their columns.

ENERGY AND TERRORISM.
With the collapse of the Soviet Union, civilization was no longer in peril from a massive exchange of nuclear weapons between two politically and ideologically opposed camps, although the nuclear weapons still remained and there are urgent steps to be taken to greatly reduce them and the potential peril of accidental or unwanted nuclear war. The focus at Erice became "Planetary Emergencies" of which energy and terrorism are the two on which I have spent most time. Richard Wilson of Harvard University has long played a leading role in both organization and substance of the energy component of the activity at Erice, and I was pleased that we were able to bring Dr. Arthur H. Rosenfeld of the Lawrence Berkeley Laboratory to

[3] International Seminar on World-Wide Implications of a Nuclear War 08/14-19/1981, pp. 192-193.

head the Permanent Monitoring Panel on Energy, now succeeded by Bill Fulkerson of the University of Tennessee.

The activity on terrorism was ably led by Ambassador Ahmad Kamal, and when I began my participation in the studies on terrorism at Erice in 2004, it was under his leadership. We soon agreed that progress could be made more effectively by doing most of the work on terrorism in two sub-PMPs. One of these dealt with Motivation, led by Lord John Alderdice of Belfast, who had played an important role in settling the conflict in Northern Ireland. I was asked to lead the sub-PMP on Motivation of Terrorist Acts. Among my most active participants were Munther Haddadin, of Jordan, and Dr. Sally Leivesley of the UK.

In 2007 the two sub-PMPs became PMPs in their own right, with the much lamented departure of Kamal. The PMP on Mitigation of Terrorist Acts has concentrated on the potential for massive death and destruction on scales thus far not seen but on the order of tens or hundreds of thousands of potential deaths. The two tools that might be used to accomplish such goals are biological agents and nuclear explosives.

I had long worked on nuclear weapons and had warned about terrorist use of nuclear and biological weapons and had participated as an author in the 2002 National Academy of Sciences book.[4] Our Mitigation PMP realized that the best way to avoid harm from terrorists would be first to motivate all people not to cause such harm to society, and, second, to intercept by intelligence and police and military action plots to use BW or nuclear weapons or materials. Nevertheless, we felt that there was much to be done in enabling populations to cope with biological of nuclear attack, and we found that particularly in the area of biological attack, it was even possible to reduce the scale of an attack with agents for a contagious disease so that the ultimate death toll might be only a few times the number of cases infected by terrorist action.

These sessions in Erice emphasized non-pharmaceutical intervention and personal protective measures (PPM). It was recognized that most of the effort involved in obtaining effective prevention would be not in defining the PPM but in providing the message, training, education, and organization for effective PPM. The work was published in the reports of the Mitigation PPM and led to a two-day workshop at Columbia University[5]. This may have influenced U.S. government activities in this area, in view of participation by the Centers for Disease Control and Prevention. But much more needs to be done.

On the nuclear threat by terrorists, not much can be done to mitigate the impact of a terrorist nuclear explosion, except to encourage people to get indoors and stay indoors for an hour or more before taking action. So the Mitigation PMP concentrated on "dirty bombs" or, more properly, "dirty dust"—the dissemination of radioactive material. Strongly believing in younger and more vigorous direction of the Mitigation PMP, I applauded the appointment as my successors of Dr. Sally Leivesley and Alan Leigh Moore as co-leaders of the Mitigation PMP.

---

[4] "Making the Nation Safer: The Role of Science and Technology in Countering Terrorism," by L.M. Branscomb (Co-chair) and R.D Klausner (Co-chair), et al, National Research Council of the National Academies, National Academy Press, Washington, DC, 2002. http://books.nap.edu/catalog/10415.html

[5] "Next Flu Pandemic:What to Do Until the Vaccine Arrives?" by S.S. Morse, R.L. Garwin, and P.J Olsiewski, *Science*, Vol. 314, pp. 929, 10 November 2006.

Here our main point was also to have a realistic view of the damage that might be done by dirty dust, to keep people inside until the situation had been surveyed, and to prescribe simple measures of cleanliness that would be largely protective of life and health in the presence of contamination.

Finally, we emphasize the need to revise the standards of environmental protection that are appropriate to guide industry and science under normal circumstances in the avoidance of even small amounts of contamination and small incidence of illness or loss of life, but that would be devastating to society if imposed after a terrorist release of radioactivity.

CONCLUSION

A brief review of the Erice volumes of 1981-1983 reporting on the "International Seminar on Nuclear War" reveals the great interest that would be associated with the presentations and particularly the informal discussions. For this, and the consequences that have flowed from these activities at Erice, we owe much to the spirit that founded and directed the activity all these years, Antonino Zichichi. Now that information technology has advanced to the point in which the World Wide Web is universally available, it should be possible to digitize the valuable books and to present them on the Erice website. The work of many would thus become much more widely available and would add to the interest and support for this work.

Bernardino Ghetti

Whilst I had known about Professor Antonino Zichichi and his Erice mission for many years, I first met Nino at the 1996 seminars. However, as I reflect back on my time involved with him, it seems that I have known him for much longer. I first learned of Nino from my father-in-law who, like Professor Zichichi, was born and raised in Trapani, which lies between the seashore and the foot of Mount Erice. My father-in-law described Nino to me with pride and admiration as a man with great scientific knowledge and vision.

I found that Nino had created an island populated by scientist heroes at the top of an ancient mountain steeped in history. There, on top of the mountain, Nino and his heroes would explore new ways to foster knowledge in order to make the world a better place. First, they wished to understand the inner mysteries of matter, discover the smallest particles and capture their energy. Nino was also able to capture the energy of man and to lead his heroes in the most diverse directions of scientific exploration. He reminds me of the words that Dante Alighieri's Ulysses says to his companions during the perilous journey:

Considerate la vostra semenza, fatti non foste a viver come bruti, ma per seguir virtute e canoscenza;"	Consider ye the seed from which ye sprang, Ye were not made to live like unto brutes, But for pursuit of virtue and of knowledge;'
Li miei compagni fec'io sì aguti, con questa orazion picciola, al cammino, che a pena poscia li avrei ritenuti;	So eager did I render my companions, With this brief exhortation, for the voyage, That then I hardly could have held them back;
e volta nostra poppa nel mattino, de' remi facemmo ali al folle volo	And having turned our stern unto the morning, We of the oars made wings for our mad flight

Nino's and his companions traveled from the physics of matter to matters of the environment thus directing his efforts to prevent human self centered behaviors leading to pollution and wastelands. From here, Nino broadened his focus and strengthened many scientists' desire to solve problems associated with human diseases devastating young, mature and aged individuals. Often, Nino has directed his attention toward the spirit, looking at God's creations in unique ways. It has been a privilege to work with Nino and to be able to contribute to the Erice community by bringing forward issues related to human brain and behavior. We may say that this was another effort by Nino to improve the human experience. Nino's critical mind and humor were always present with us and his influence made us see things from an alternate viewpoint as a physicist looks through a different window from that of a physician.

In Erice, we were able to discuss and reflect on the illnesses that rob the human spirit of its most precious gifts. We had beautiful opportunities to bring together modern day heroes from the neurosciences to explore the molecules of the human brain and how those can change toward abnormal states bringing forth diseases of the brain and altering the qualities that distinguish us as humans.
From the mountain of Erice, 3000 years have yielded a spring of spiritual enlightenment and knowledge ...as Heraclitus said ”παντα ρει και ουδεν μενει” “everything is in flux and nothing abides”.

Alberto González Pozo

Since 2000, I have been invited to Erice by the World Federation of Scientists and the "Ettore Marjorana" Centre for Scientific Culture, to join the *International Seminars on Nuclear War and Planetary Emergencies* that take place there. Being an architect and planner, specialized in urban problems and cultural heritage preservation, I was invited to participate in the group working on Limits to Development issues, with emphasis on Megacities. Accordingly, my first contributions to those meetings dealt with Urban Growth, Water Supply, Wastewater Disposal and Urban Mobility in Mexico City, one the biggest settlements in the World. Gradually, the interest of the group turned to other topics dealing with Sustainability, like Migration and Climate Change. Therefore, I had to focus on these interesting fields of knowledge too, showing to my colleagues the aspects of the situation in my own country and region .

After the first years of my visits to the Erice Seminars, I realized that not all the participants were scientists or technologists. Some, like me, work as scholars in universities or in problem-solving teams in public or private organizations. There are engineers, physicians, experts in foreign affairs as well as several distinguished journalists. This diversity of profiles, backgrounds and values among the participants gives the sessions a lively and interesting character.

It has been a pleasant experience to come each year for a week to Erice and find there not only the half-dozen colleagues in our Limits to Development group coming from Europe, Africa, Asia and the Americas, but also to listen to more than a hundred other experts from all over the World. All bring interesting presentations about questions where Science and Technology have found new answers to the most important problems that mankind faces today, called *Planetary Emergencies*. Each scholar is probably among the best available in his field, and all of them show the latest version of the "state of the art" in their respective areas of knowledge.

Our Chairman, Professor Antonino Zichichi, who is at the same time the founder and main promoter of the Seminars, recalls in the opening presentation each year the interdisciplinary character of the meetings and shows the way he devised to give some order to the discussions. He looks at the Planetary Emergencies as a tree with fifteen main branches, where each one lets grow smaller branches. Not all of the themes and subthemes are considered yearly, but some are regularly monitored. This is the case of themes 1 to 5 that deal with *Water, Soil, Food, Energy* and *Pollution*: all of them crucial for the well-being of many countries and regions in the world. Then comes theme 6: *Limits of Development*, where I participate, with three subthemes: Demographic Growth, Megacities and Resources Provision, where emphasis is put on sustainability. Theme 7 is closely related to the latter, because it is focused on *Climatic Change* that each year grows in importance because it is linked to the survival of life on our planet. Themes 8 to 10 have rather apocalyptical names: *Global Monitoring of the Planet, New Military Threats* and *Science and Technology for Developing Countries to Avoid North-South Environmental Holocaust*, where the presentations deal with the assessments of the threats they represent and the ways to overcome them

Themes 11 and 12 are related to health-care emergencies like *Problems of Organ Substitutions* and *Infectious Diseases*. During several years, for instance, issues such as AIDS, SARS, Avian Influenza or Spongiform Encelophaty have been clearly exposed by experts, some of them of Nobel Prize rank.

Theme 13 is probably the most humanistic and, at the same time, the most ideological of all, too. It deals with *Cultural Pollution,* but was divided in the following subthemes: Political Violence (including terrorism, for instance), Education, the Values of Science, Science and Faith, Science & Technology and Media, and finally Modern Technology and Democracy.

Themes 14 and 15 have an apocalyptical content, too, because they deal with *Common Responses Against Cosmic Objects* that may hit our planet and *Huge Military Investments*, that can trigger new regional conflicts with more destructive weapons, or even to a world war and a nuclear holocaust that scientists must do their best to avoid.

Even if the field of knowledge of many of these themes and subthemes may be far away from the one that most of the participants are working on, for all of us it is of great interest to listen to fascinating presentations of recent progress in each topic, as well as discussions that frequently arise around the premises, the sources of information, or the methodology employed in their assessment. Thus, the typical seminar week in Erice represents between 50 to 60 presentations in general sessions, besides 10 or more in each workshop within the fifteen groups.

Fortunately, the infrastructure of the Seminars is excellent. Erice itself, built and rebuilt successively by Greeks, Romans, Arabs and Normans is now a wonderful example of a small Medieval town carefully preserved, where an occasional walk looking at its picturesque narrow alleys and the old monuments is worthwhile. A former Dominican church is now the main auditorium of the meetings; and two other former religious buildings (a small convent, San Rocco, and a bigger cloister, San Francesco), properly restored and adapted, serve as headquarters of the "Ettore Marjorana" Centre (with more meeting facilities), and as a hostel and social meeting place, respectively. Most of the participants are hosted in other city hotels and the mild summer climate of Erice (750 meters above sea level) is very pleasant.

Everyone forgets completely his or her car because all the places can be reached easily at walking distance, plus there is the advantage of the free public transportation service that runs every fifteen minutes around the city in small vehicles. The excellent Sicilian cuisine is enjoyed at lunch in *trattorias* where small groups extend the discussions of the morning sessions and have the opportunity to get to know each other better. And every night, there is a dinner where all the participants (many of them together with their spouses) can interact personally with the rest, making new and interesting acquaintances. An afternoon during the week is free of meetings, and all are invited to enjoy a short tour organized to different interesting places in the western side of Sicily, like the ruins of the Greek temple and the neighbouring theater at Segesta, the old towns of Trapani or Mazzara del Vallo, and other places of touristic value.

All these services are organized by a small but very efficient group composed by the World Federation of Scientists and the "Ettore Marjorana" Centre staff. They are most helpful with every problem that may arise and always find a solution to practical questions.

Professor Zichichi and his team manage to invite new distinguished scientists each year, as well as other important personalities of the political, cultural and even religious panorama. Professor Zichichi is convinced that they must be acquainted with the latest scientific findings in order to take better decisions in development issues.

All of us that have enjoyed the opportunity to come to the Erice Seminars accept with gratitude and appreciate the great effort that Professor Antonino Zichichi has done in promoting and organizing the sessions for almost three decades. I add here my sincere recognition to the role he has played in the advancement of science and technology through what I now call *the Erice Experience*, an interdisciplinary intellectual adventure of the greatest importance for the understanding between scientists from very different countries and cultures. Regardless of our particular research field, we all celebrate his important role promoting new ways of scientific progress through international exchange.

Brian Heap

My first experience of these important seminars occurred on 19-24 August 2006 at the generous invitation of Professor Zichichi who, as a colleague on the NATO Science Committee, had long invited me to attend. This was the first time that it was possible and the effect was dramatic, not only because of the ambience of the place and participants but because of the impact of the programme, one of those rare events when content shakes you to the core. The topic in question was plastics and their impact on the environment. Was this a planetary emergency?

My own research interests in sustainable consumption and production proved to resonate closely, and Professor Zichichi's seminar provided a serious wake-up call into the matter of how we might consume and produce plastics in a more sustainable way.

Among the points that hit me were the following in summary form:

- Plastics have made a singularly important and valuable contribution to modern life particularly but not exclusively in a consumer society. The industry has increased production from 1960 to 2000 from 0.5m tons pa to 25 m tons pa, but the amount regenerated has increased from zero to 1.5 m tons pa. The gap between production and regeneration is growing and plastic litter in the USA has now reached a figure of 60-80% municipal waste stream. The rate of rise has even taken the industry by surprise.

- The development of biodegradable plastics has been an important step but research shows that much of the debris blocks drains and rivers raising the question of the definition of biodegradable.

- Plastic piping now replaces old pipes to deliver potable water. Phthalates are used as a class of chemicals in a wide range of products including flexible plastics, industrial solvents, personal care products and medical devices. However, at high dosages they are anti-androgenic and disrupt normal male reproductive tract development in rodents.

- Bisphenol A (BPA), an oestrogen-mimicking chemical, is now produced in excess of 6bn pounds (weight) per year and is used to create polycarbonate plastic for lining metal cans and for many other products. Over the past three decades close attention has been given to whether such chemicals leach into the environment and/or into food and beverages, and whether they could act as endocrine disrupters. I had understood that the risks were low but recent work in laboratory animals has raised questions about their potential health effects particularly in relation to insulin secretion and insulin resistance typically found in diabetes and obesity.

- Significant levels of phthalates and BPA have been found in pregnant women and fetuses and the issue is whether they are within tolerable daily levels of intake, or a matter of concern because insufficient attention has been given to the discovery of alternative additives.

Arising from the Erice seminar many came away wanting answers to further questions. Could a better product-recycling loop be developed to effectively reduce the plastics burden on the environment? Is there a scientific consensus about the increased risks associated with the rising accumulation and deposition of multiple forms of plastics and additives? What policies could be constructed to guide the public to conserve plastic and re-use? Is there an additive effect of exposure to multiple plastics and their leached additives? Could an international partnership be forged between scientists, policy makers and industry to examine corporate social responsibility along the lines of the effect of CFCs on the ozone hole?

Some of these questions have answers from present knowledge and I am pleased to report that a relevant issue of the *Philosophical Transactions of the Royal Society* will soon be published. This volume will contain a range of informative papers about these issues. Clearly, opportunities and rewards exist for those committed to sustainable consumption and production as a strategy for the future. The eco-opportunities presented to industry by the world market for recycling, waste management, pollution control and energy efficiency has been estimated to be more than £600 bn pa, competing strongly with global aerospace, car and chemical industries. Cornucopians will claim that technology-fixes resolve impending crises whereas catastrophists question whether the fixes will deliver in time. The 850m long-established consumers in rich nations are already being joined by an even larger number of new consumers in 20 developing and transition nations. A message for today, therefore, is that time is not on the side of business people and policy makers who are only just beginning to think about sustainable consumption and production. However, it remains an optional strategy waiting to be adopted not least in a time of recession. Science and technology have a crucial part to play but dealing with overconsumption in the North and underconsumption in the South will not be resolved by science and technology alone. Creating a sustainable future demands a deeper understanding of environmental quality, social equity, economic development and the determinants of consumer behaviour.

If the answers to the questions raised above can be answered satisfactorily by full and open disclosures it will be a tribute to the Erice International seminars that this topic need not become a Planetary Emergency. If the answers do not satisfy, then the continuing debate must clarify what preventative action is needed by scientists, industry and the consumer.

Walter F. Huebner

The threat of a collision of an asteroid or a comet nucleus with Earth is a worldwide hazard that was recognized several decades ago. Such collisions have occurred since the formation of the Earth and the planets in general and they continue today. In July 1994, the world witnessed the collision of Comet Shoemaker-Levy 9 with Jupiter. The nucleus broke into at least twenty-one large pieces with diameters up to 2 km, many of the pieces created fireballs in Jupiter's atmosphere as large as the Earth itself. About one hundred years ago an asteroid or comet nucleus exploded in the atmosphere above a very sparsely populated region of Siberia, the Tunguska event. It flattened about 2,000 km^2 of forestland. While the collision frequency with Earth has subsided as the supply of asteroids and comet nuclei in the planetary regions dwindles, the supply is far from being exhausted and collisions of asteroids, meteoroids, and comet nuclei with planets continue. Table 1 summarizes some better-known impacts on Earth and airbursts and fireballs in the Earth's atmosphere over the last one hundred years. As stated in the 2006 report of the task force on potentially hazardous near-Earth objects (NEOs), about thirty impacts are recorded per decade.

Although the objects listed in Table 1 are small objects, most of them only of very little danger regionally, the Tunguska event in Siberia and the collision of Comet Shoemaker-Levy 9 with Jupiter in 1994 are reminders and warning signals that we should take seriously. It is only a matter of time until a larger object may collide with earth. The asteroid 99942 Apophis will pass through the geo-stationary orbit of Earth in 2029 where weather, communication, and other satellites circle our planet. Impacts of objects larger than about 2 km in size would threaten the extinction of many species and indeed threaten the existence of mankind. Such impacts are extremely rare. The Tunguska object is thought to have been about 50 m in diameter and caused significant local damage. Impacts by such objects can be expected to occur every few hundred years. Statistically, at least one small celestial object in the 10-meter size range passes Earth within Moon distance daily. The kinetic energy of such an object, depending on its density and relative speed, approaches the equivalent of about 1 megaton of TNT. Only the upper atmosphere of Earth offers any measure of protection from the direct effects of a sudden energy deposition from objects on a collision course with Earth. This is clear when we examine bodies not protected by an atmosphere, such as the Moon and Mars.

The size distribution of NEOs suggests that small objects will be encountered most frequently (a Tunguska-sized object about once in 200 to 1000 years). 50m objects are at least 100 times more abundant than 1 km sized objects, and are about ten times more abundant than 140 m objects. Since small objects tend to be very faint, they may be close to Earth when they are found and warning times for an impending collision may be very short.

Table 1. Some Recent Meteorite Impacts, Air Bursts, and Fire Balls

Date dd/mm/yyyy	Place	Yield* TNT	Size	Comments
30/06/1908	Tunguska, Siberia, Russia	10-15 Mt	50-60 m	Tunguska event
28/06/1911	Nakhla, Abu Hommos, Egypt		10 kg	
13/08/1930	Curuçá, Amazonas, Brazil	0.1-1 Mt	10-50 m	Curuçá event: air burst. Significant ground damage.
11/12/1935	British Guyana			
12/02/1947	Sikhote-Alin, Primorye, Russia		10^5 kg	Iron, type IIAB, about 100 craters, largest: 26 m across, 6 m deep
30/11/1954	Sylacauga, AL, USA		4 kg	Stone Chondrite
07/04/1959	Pribram (Freiberg), Czech Rep.			Ordinary Chondrite found: 9.5 kg
31/03/1965	Revelstoke, BC, Canada	20 kt		Carbonaceous Chondrite
31/05/1965	SE Canada	0.6 kt		1 g meteorite found
17/09/1967	Vilna, Alberta, Canada	0.6 kt		2 small fragments found.
22/09/1979	Southern Indian Ocean	2 kt		
09/10/1992	Peekskill, NY, USA			Chondrite found: 12.4 kg
/04/1993	Dubbo, Australia			
18/01/2000	Whitehorse, Yukon, Canada	480 t	4.6 m 1.8%10^5 kg	
06/06/2002	Mediterranean Sea	26 kt		
25/09/2002	Bedaub, Irkutsk Oblast, Russia	0.5-5 kt		Bodaybo or Vitim event
07/06/2006	Reisadalen, North Troms, Norway	0.1-0.5 kt		
15/09/2007	Carancas, Puno, Peru	0.005 kt	3 m	Carancas event. Chondritic meteorite. Crater: 13.5 m across, 4.5 m deep.
10/07/2008	Northern Sudan	1-2 kt		Bright fireball seen as far as Europe. First fireball predicted in advance.
10/05/2007	Puerto Lápice, Spain			About 280 g of eucrite found.
19/11/2008	Edmonton, Alberta, Canada			

*) The equivalent yield of 1 ton (= 1 Mg) of TNT = 4.184×10^9 J.

For the first time in man's history, we believe that countermeasures can be undertaken to avoid such collisions. Professor Antonino Zichichi recognized this as a major world emergency and created the Permanent Monitoring Panel for Cosmic Objects (PMPCO) as part of the International Seminars on Planetary Emergencies to study and advise appropriate authorities about potential mitigations and countermeasures.

The first chairman of the PMPCO was Prof. J. Mayo Greenberg from the University of Leiden, The Netherlands. The first sessions summarized work that had been done in the USA and other places, but with more emphasis on comet nuclei.

Mayo Greenberg organized the first meeting I attended at the Ettore Majorana International Foundation and Centre for Scientific Culture in Erice in August 1998. We discussed NEOs in a small group that included Jondale Solem (Los Alamos), Jack Hills (Los Alamos), Bill Boetke (Southwest Research Institute), and John Remo (Harvard Smithsonian Center for Astrophysics). The meeting concentrated on the need to determine internal geophysical and geological properties of NEOs. Continued discussions led to the decision to have a larger meeting after the international asteroids meeting in Palermo in 2001. For this purpose Mayo Greenberg conducted the 6th Summer School on Space Chemistry in Erice.

Mayo Greenberg attended the Summer School only for one week because he was already sick with cancer of the pancreas. Together with Alberto Cellino (Torino Astronomical Observatory), I conducted most of the sessions. The topic was the "Properties of NEOs: Know Your Enemy." Many Japanese colleagues including Akira Kouchi (Hokkaido University), and Tetsuo Yamamoto (Nagoya University), attended with their students. We took a tour to one of the destroyed ancient temples. Stones from the destroyed temple were strewn all over the place. An analogy to the "rubble pile" structure of asteroids was formulated in jest. We identified many properties that will be required to counter any threatening object. However, four groups of data that will be needed most critically of any potentially hazardous NEO are: (1) its mass, (2) its mass distribution, (3) its material strengths, and (4) its internal structure. Global (whole body) properties, such as material strengths and internal structure, can be determined best from the analyses of permeating waves: electromagnetic (radio) and sound (seismic). Summaries of the workshop were published in International Seminar on Nuclear War and Planetary Emergencies. An abbreviated version was published in the journal for Meteoritics and Planetary Science.

At the Seminars on Planetary Emergencies meeting in 2001 Mayo Greenberg had planned to present a summary of our summer school and workshop on NEO properties that we held two months earlier in Erice. Because Mayo was sick with cancer, he did not attend. I was prepared to give a summary, but the morning that Mayo was scheduled, his contribution had been taken off the agenda. This was a great disappointment, but Prof. Zichichi invited me to present an oral summary to him in the presence of Italian journalists. A summary of the discussions with questions and answers appeared in the Italian press.

At the Seminars meetings in the following years, we continued to explore and develop plans for mitigation of potentially hazardous objects. Various experts were invited to our PMP for Cosmic Objects. Among them were Don Yeomans (Jet Propulsion Laboratory), Russell Schweickart (B612 Foundation) who presented concepts about the Gravity Tractor; Hajime Yano (Japan Aerospace Exploration Agency) who presented plans and new results from the Hayabusa mission to asteroid Itokawa, Alan W. Harris (German Aerospace, Berlin) described ESA's plans for the Don Quijote mission. We also obtained written input from Anatoly

Zaitsev (Lavochkin Association) about the Conceptual Project of the "Citadel" Planetary Defense System planned in Russia.

Interactions with European, Russian, and Japanese colleagues led to enhancing cooperation and coordination on an international scale among different national and regional organizations on research and technology development for disaster mitigation of PHOs. We also started to develop a matrix for options according to whether the threatening object is an asteroid or comet nucleus, its size, and the warning time. Options may range from benign methods of deflection to aggressive methods of destruction (but avoiding catastrophic disruption) and everything in between (e.g., ballistic impacts, chemical and nuclear explosive, gravity tractor, etc.). For each method we discussed what technology is needed to carry out the required mitigation, and the potential consequences. These discussions included political issues for mitigation. Roland Timerbaev (former Russian ambassador to the UN) helped to clarify some of these issues.

All of these discussions to which many more participants contributed than we have space here to describe, led to the formulation of plans for comprehensive countermeasures against PHOs as we outline next.

While deflection techniques will always be preferred, we will also consider options for destruction of a PHO in emergency situations, such as very short warning times, which means very short times for reaction. In all cases we will investigate the threshold where any aggressive technique might lead to unacceptably large fragments remaining on a collision course with Earth. To avoid domestic and international political objections for the use of nuclear demolition munitions, we will explore the limits on size, composition, structure, etc. for using conventional explosives on PHOs.

The options for countermeasures proposed to be studied include conventional and nuclear explosives either to deflect a PHO from the collision path with Earth or destroy it. The potential for deflection rests on using conventional explosives to produce explosives gases and ejecta that impart momentum to the object to alter its trajectory. Our proposed project does not conflict with other possible projects for deflection of PHOs with long warning times, such as the "slow push alternatives." It complements them. For example, if a slow push method should fail during the many years of continuous operation needed to affect the mitigation, our methods can come to the rescue. Alternatively, they can be used in place of slow push methods and may be more cost effective.

The question about which method to apply can be analyzed by examining the engagement space. The engagement space is multi-dimensional and depends on many variables. The dimensions include the reaction time, the size of the threatening object, its chemical composition and physical structure, the type, intend, and magnitude of the countermeasures, etc. If the reaction time is short, then aggressive methods must be employed. If the object is massive, then more powerful methods must be applied. Different methods should be applied if he object is monolithic or a rubble pile. The intended countermeasure can be deflection, which is the preferred method, or destruction and dispersal of the largest fragments. As for the type of countermeasures, they can, for example, be a hypervelocity impactor, conventional explosives, or nuclear explosives. Explosive devices can be applied in a sand-off, surface, or subsurface explosion. For political reasons, we propose conventional explosives for small objects with short warning times.

Our plan also includes testing of simulated materials. Asteroids and comet nuclei are composed of materials of low density. We must make sure that the proposed methods are effective when applied to such materials. To obtain more relevant data we propose seismological characterization of asteroids. A database of geophysical and geological properties of asteroids and comet nuclei has been set up, but needs more data. We will also design, build, and test relevant components for applying countermeasures. We suggest that an appropriately equipped spacecraft observe the effectiveness of countermeasures and record the results.

Our plan calls for different types of countermeasures depending on the situation encountered. Thus flexibility is of great importance. Since we also want to counter short warning time, i.e., reaction time objects, readiness is very important. Thus launch capabilities and launch sites must always be at the ready. One possibility is to use some missile silos that are now being decommissioned. Nuclear munitions are by far the most efficient in terms of yield per unit mass launched and are technically mature. Being able to react on short warning times, our methods can also be applied to long-period comets that are generally not considered. Another important aspect is coordination of plans. This is an international problem and needs international cooperation in tasks and projects. The Permanent Monitoring Panel for Cosmic Objects (PMPCO) associated with the International Seminars on Planetary Emergencies play a very important role in this.

Once you face in the right direction, all you have to do to reach your goal is move forward. Professor Antonino Zichichi helped us to face in the right direction.

Hisham Khatib

With fifty years involvement in science and engagement in professional life, I have participated in more than my fair share of conferences, seminars and other professional encounters. Therefore I am well placed to declare that the Erice International Seminars are definitely among the most inspiring of these. This is due to the variety of topics, the high caliber of the participants and their quality contributions, the streamlined organization of the seminars and last but not least the captivating location of the City of Erice.

Although the Erice International Seminars commenced with emphasis on nuclear physics they have extended to encompass most aspects relating to science and quality of life including – besides nuclear energy and proliferation – sustainable development, human health, the weather and the environment, frontiers of science, cyber security, and overcoming terrorism; as well as other challenges facing humanity. Such quality variation enriched the seminars, increased and enhanced its benefits and widened participation.

The Seminar is practically two seminars, the official Seminar taking place in the comfortable surroundings of "E. Majorana" Centre for Scientific Culture with its panoramic view of the Mediterranean, and the other is the daily meetings and encounters over dinners (and lunches) between participants where acquaintances are built, ideas exchanged and life long professional and personal relationships are established. These led, in my case, to continuous exchange of ideas, papers and other information with most rewarding results.

Erice Seminars are also the site of brains encounters from the East and the West, and the meeting place of scientists from developed and developing countries. Scientists from at least fifty countries have already participated in the Seminars. This is a meaningful effort in engaging people from different nationalities, varying backgrounds and cultures to share scientific knowledge and achievement for the benefit of all.

Through my involvement with the World Energy Council (WEC) which is the world's leading non-governmental energy organization, I have tried to establish a working relationship between the Erice Seminars and the WEC. This led in 2005 to a well attended joint workshop on clean use of fossil fuels and Carbon Capture and Sequestration (CCS). The results of this meeting were not restricted to the Erice Seminars, but went beyond that to the benefit of the wide membership of the WEC. I am hoping that more of this kind of collaboration will take place in the years to come so that the Erice Seminars spirit and outcome will benefit a wider audience.

By Erice standards I am a relative newcomer. While the activities of the Seminar extend for over 27 years, my participation so far extends to seven annual seminars all of which took place in recent years. Therefore, I did not have the pleasure of meeting some of the earlier generation of distinguished scientists who started their participation in the Seminar in its earlier years – Paul Dirac, Edward Teller and Eugene Wigner, to mention only a few. During my recent participations I was personally impressed by the Chairmanship of Professor Antonino Zichichi, whose professional expertise as a scientist is worldwide recognized. However, I was witness to other distinct aspects of his character – immense energy, good humor, personal knowledge of all participants and their names, his relaxed manners, which very much enriched the Seminars. I was very pleasantly surprised to realize that Professor Zichichi's reputation extends over even to Jordan. Not only is he well known to the Italian diplomatic core in Amman but also to all Italians living in Jordan as well as Jordanians well acquainted with Italy. Apparently his Sunday science show on Italian TV helped in that. He is now my "unforgettable character".

One of the most capitulating aspects of the Erice Seminars is the City of Erice itself. I come from the Holy Land and had my school education and upbringing in Jerusalem. The common aspects and similarities between Erice and Jerusalem are most striking – mountainous location, architecture and buildings, local stone structures and stone paved roads, small indoor gardens and small shops emphasis on local souvenirs. Over the years I have developed a special relationship with the place, as much as with the International Seminars.

Vasiliy Krivokhizha

To speak adequately about Erice is a challenge for a number of reasons. To start with, it is a challenge because of large scale, literally multidimensional meaning of the very place and research center which has become during last few decades one of the symbols of current global science. First is the challenge to your intellect to accentuate the actual in the description of the whole in combination with personal perception. In a political sense, one is challenged to find an optimum viewpoint for covering various aspects of the results of science applied to human welfare during all these years by the Erice Seminars. Some could give the evidence of themselves years later and not necessarily in direct way.

Moreover, keeping in mind the original scientific ideas behind the international and Italian efforts to establish "Ettore Majorana" Center for Scientific Culture, the concrete aims of signers of the Erice Statement, as well as the wide scope of initial political aspirations of the organizers and participant countries during tensions of the Cold War period renders evaluating the situation in its development, many times more difficult.

Also it must be quite clear from the beginning that no institution can develop successfully throughout a nearly 50 year history jubilee without significant transformation, including some organizational principles, to remain responsive to the dynamics of the global international situation.

Nevertheless, the aforementioned challenges are far from exhaustive. One might also, for example, compare scientific process at Erice with organization of work, seminars, conferences, publications, etc. at few other scientific centers throughout the world.

A far more complicated matter is to cover the specific features of Erice that derive from the place itself. Envision the surrounding world from the cliff, with ever changing vistas of clouds beneath and simply as a vertical projection somewhere in the skies – a perspective that is more familiar to those who come to Erice from the plains. These aspects influence the very mood and approach (scientific, philosophical, and political – at least to the extent) of identifying and solving high priority problems over any concrete period of time. The whole surrounding atmosphere makes people more receptive to each other's thoughts and more tolerant. Erice is a transformative place in that sense.

Simultaneously, the very fact that during at least twenty five centuries Erice remained one of the centers of successive dominant civilizations proves the very special character of the town and place of Erice – its basic internal stability (its strong "spine") in the midst of a world of uprising and declining societies and cultures. The realization of this rare phenomenon is one of the first and brightest impressions of Erice.

The eternal character that constitutes the outwardly appeasing appearance of Erice does not, however, necessarily imply a philosophical, pacifying mood of discussions during the Seminars. On the contrary the clash of opinions may be accompanied by quite intensive discussions as have taken place around the principle position of participants concerning such subjects such as the expansion of NATO, Chechen affairs, the contradiction of state interests surrounding the third US missile deployment area in Europe, or divergent evaluations of the circumstances of the war in the Caucuses last August and presently independent status of South Ossetia.

From the viewpoint of the "Erice" process of communications among participants in "the quest for truth", I would like to point out that many people highly appreciate the free manner of presentation of opinions and very character of discussion. This algorithm of the working sessions is very useful for a number of reasons. One may discover better understanding of the usually wide spectrum of positions on any given acute problem when eloquence does not hide much of the initial aspirations and interests behind the ideas and suggestions of the sides involved. For some participants, a further attractive feature is a trend to deliberate on social, political aspects of what seem at first glance primarily technical results of research works prepared.

The work of the Mitigation PMP illustrates this feature well. Targeting the achievement of applied recommendations is one of the brightest characteristics of the Erice School. Yes, there is a basis to speak about a School of Erice. The twentieth century has been marked by distinguished results in different spheres of sciences. There are centers famous for knowledge and results in their respective realms of sciences. But this process of focused research sometimes is short of the potential to produce a broader human vision and consequent recommendations in the form of systemic knowledge, combining multidisciplinary approaches.

Today under the pressuring realities (differing for different countries) of the global financial and economic crisis, it is more and more evident that the original problem of cooperation in coping with the results of the crisis roots derives not only from competitive relationships among countries, but also from the fragmentation of economic knowledge (Why? – It's another question). If you ask your friend what a school of economy is better now? The answer usually sounds like – What sphere of economy do you have in mind? What kind of systemic multilateral efforts may be undertaken when global recommendations are based on individual visions of countries? Partial knowledge of the situation, where there is no basis even to correlate interests and measures and when even the key players of world economy speculate about some prescriptions that are probable and acceptable for them, betrays the weakness of unilateral instruments. All this demonstrates the distinction between the functional potential of even prominent scientific centers and the Schools of contemporary thought and necessity for a comprehensive, widely accepted picture of current situation. And it makes the difference…

Even speaking in the style of "Song of Songs" one would exaggerate to say that the Erice process –over the years – has succeeded completely in recommending solutions to the most strident international problems of the late Cold War period and after the period following the disintegration of the USSR with the public frame of mind on bent reconciliation and a new Golden Age, in accordance with the slogan which temporary prevailed, about absence of winners and losers in the Cold War. The realities of politics turned out to be different.

Yet it is important to keep in mind that the initial purposes behind EMFCSC remain valid as do the positions of the Erice Statement. In practice, of course, most of ideas and proposals deserve rethinking and specification in light of the current political cultural international environment: ranging from the meaning and interpretation of doing "only pure science" to the proposal close in content and context "to prevent the acquisition of nuclear weapons by additional nations or non-national groups." Even neutral and righteous proposals about necessity of efforts "to reduce or eliminate restrictions on the free flow of information, ideas and peoples" sounds – twenty years after the Cold War – in many aspects dubious in dealing by itself with the central problem that is very difficult to cope with decisively.

One may pose a question closer to the members of Panel on the Problems of Terrorism about correlation of priorities of problems related to international terrorism, role of "pure science" in the process of deliberation of methods and instruments to solve these problems on the maximum possible practical level, and the influence on international cooperation in this sphere by taking thoroughly orchestrated competitive political campaigns among countries involved in the process. Speaking more or less objectively, it is not a failing but rather a challenge for the Erice School to operate in such a situation. And whatever the actual results – primarily of political nature – the implementation of original ideas has come a long way in facing what are still priority problems.

The Schools of Erice deserve special notice as they are actively involved in the maturation and educating process of young scientists for whom to understand each other's deeds and motivations is easier than for people of previous generations. It makes sense to repeat also that the most simplistic way to produce quickly planned results – to demonstrate unanimity and possibility to come to common denominators and recommendations – is to take a bunch of soul-mates, as simple as that. This practice is well known, and one may find many examples of it. Moreover, projects of this type frequently receive good financial support. However, they may have not much in common with science and positively do not contribute to the stability of interstate relations. It is not a way of Erice.

To make a long story short concerning the role of Erice in contemporary scientific culture, as my personal touch, I would like to give a small illustration. Once upon a time (in the early fifties) a friend of mine, being a student, obtained the possibility to advance his Chinese language skills in Beijing. Under the strong influence of the ancient culture of the country he came to the conclusion that buying during his stay the same sort of tea at the same good, old tea-shop might be considered by the people around him as a striking illustration of his deep understanding of roots in Chinese culture. Once he witnessed the talk between the owner of the shop and his old client. The latter said that during decades he was buying very special sort of tea from the northern slope of one locality famous for its tea hill. But by mistake he was given previously by one of the youngest salesmen (still more than sixty years old) the tea from the southern part of the aforementioned hill. The owner apologized, then kept a long interval of silence, mostly to accentuate his astonishment with very fact that even youngest of his workers could make such a mistake. He could understand and accept the explanation that under the specific weather conditions of that year "the young guy" could be confused by the rather similar aromas of teas from northern and eastern slopes. Nevertheless, it was beyond his imagination to realize the origin of mistake between the teas (of the same kind) from northern and between southern parts of the hill. For him as well as for client these teas were fundamentally different.

This example offered me few lessons. One of them was to feel the distinction between the people, scientific, analytical centers qualified to solve problems in narrow spheres of science and life (may be very successfully), and those who are capable of seeing the whole picture with both the retrospect of its development and the strategic perspective simultaneously, i.e. in the context of particular Scientific Culture. Centers such as the Majorana Center of Scientific Culture at Erice belong to the small group of – I would say – lighthouses, which give us guidance points in our diversified and complicated world. This is a high characteristic which should be, probably, the prime and central criterion of their value.

Axel Lehmann

On June 2nd, 2000, I received an email from my colleague Prof. Dr. Reiner Huber asking me to give a talk on security threats arising from digital stored and transmitted data at an "International Symposium" and a co-located session of a Working Group on "Missile Proliferation and Defence". The Symposium and workshop organized by the World Federation of Scientists would be held in Erice, Sicily. Reiner Huber knew that I had been recently involved in a research study on the subject of "information security threats" from my perspective as computer scientist. Indeed, I just had finished some research on methods and tools threatening privacy, security and authenticity of digitized data, or information. In my research I was focusing on steganography - also known in literature as the method of "hidden writing" - in context of digital data storage and transmission. Prof. Huber who had attended these Annual International Seminars on Nuclear War and Planetary Emergencies already for several years explained on my request that - based on a proposal of Russian colleagues - this years Working Group meeting "Missile Proliferation and Defence" should be directed towards "... problems of maintenance of international information security, and the role UN and other international organizations should consider. The meeting should be devoted to the problems connected with information weapons and the danger of information wars, information terrorism, and information crime ...". He mentioned these WFS meetings would also address other severe planetary emergencies, always from an interdisciplinary perspective, like prevention of nuclear war, or threats arising from limited water and energy reserves in the context of a rapidly increasing world population. He added in his response that besides the scope on planetary emergencies, he always enjoyed the interdisciplinarity and openness of these meetings reflected by the variety of scientists attending from all over the world - a unique meeting place for scientists from different cultural backgrounds and disciplines in a historical environment.

As a computer scientist involved in research of innovative computer structures and novel applications, I was somewhat surprised, or even irritated, to consider digital information as a potential "weapon", or as a major means for an "information war". Nevertheless, I agreed, and proposed a talk on "Threats to Information System Security by (Computer-Based) Information Hiding Techniques". Some weeks later – I received a very kind and generous invitation to attend the 25th Session of the Erice Seminars on Nuclear War and Planetary Emergencies in August 2000, and to give a talk in one of the plenary sessions. As spouses were also welcome and invited, my wife Barbara, speaking almost fluently Italian, joined me.

Already in the first hours after our arrival in Erice, we were overwhelmed by the flair and spirit of Erice: by Erice's "Piazza", the historical buildings of churches and monasteries, and by the Ettore Majorana Center in San Rocco – an old monastery perfectly renovated and providing all the services and infrastructures required for holding seminars and meetings. The evening of our arrival we were asked to attend a private dinner with the host and General Chair of the Erice Seminars, Professor Zichichi, who was not known to us up until this day. Both of us will never forget this first meeting and dinner with Antonino Zichichi: We met in the restaurant "La Pentolaccia", together with a group of about a dozen scientists, 2 nobel prize winners (Prof. Lee and Prof. Siegbahn) among them. We were impressed by the hospitality and informality of our host and the attending colleagues, as well as by the openness of discussions during dinner.

The following days in Erice kept us busy and we learned to enjoy the uniqueness of these Erice Seminars. I attended the plenary sessions of the Annual Seminar and learned to enjoy the unique atmosphere of interdisciplinary presentations and discussions in this wonderfully renovated San Domenico hall. Also the evenings offered excellent opportunities to dine and discuss together with colleagues and their spouses in the restaurants close to the Center, or in the old "Marsala" cellar in San Rocco. During my academic career I have had many opportunities to attend international workshops, symposia and conferences around Europe, in North America and in Asia. But never before I had attended an event in such a fantastic and open-minded, and multidisciplinary atmosphere: Bringing together well-known scientists from different disciplines, cultures, religions and countries around the world to forward their expertise in addressing severe emergencies for our planet.

As a result of this Working Group Meeting on "Missile Proliferation and Defence", several weeks later Ambassador Dr. Henning Wegener, Vitaly Tsygichko, Dr. Andrey Krutskikh, Professor Dr. William Barletta and myself forwarded a proposal to WFS to consider the formation of a new Permanent Monitoring Panel on "Information Security" under the Chair of Ambassador Wegener. In August 2001, during the 34th Session of the International Seminar on Nuclear War and Planetary Emergencies in Erice, the first "Information Security" PMP-Meeting was organized. In the following years, the PMP was joined by Dr. Jody Westby, Olivia Bosch, Ambassador Ahmad Kamal and Dr. Vladimir Britkov, and has now expanded its membership even more (see the PMP web site: www.unib.de/infosecur). Over the following years, its members contributed in various forms to outreach activities, for example, by forwarding several special reports and recommendations to the UN, EU, and other international organizations, or by presenting papers at international conferences. The PMP and its outreach activities always benefits from its multidisciplinary membership uniting diplomats, lawyers, physicists, engineers, military experts, and computer scientists from different countries, with their concerns and views in regard to information security in a global world.

Today – after almost nine years – the PMP has enlarged its membership to almost 20 members and associate members. It has also expanded its scope of work towards cyber conflicts, cyber crime, and cyber war. As far as I know, there exists no such multidisciplinary consortium or group of scientists around the world comparable to this Panel focusing on common efforts towards information security challenges – thanks to the permanent support of Professor Dr. Antonino Zichichi and the World Federation of Scientists.

Michael MacCracken

The 1980s were a very tense period internationally—the Cold War was very cold indeed. For more than twenty-five years, peace had been maintained by the superpowers threatening each other with mass destruction. While the standoff seemed to be holding, several new factors were threatening to upset the balance. Each superpower was testing the other's tolerance by confronting each other through conflicts of third party nations. As this took place, primarily as a result of provocations by their own governments, public opinions in each nation were becoming more acrimonious, seemingly to ignore the level of civility appropriate given the destructive potential of the large numbers of nuclear weapons on each side.

Science was also adding to the tensions. The possibility of defensive weapons was being investigated, a step seen by many as upsetting the long-standing nuclear balance. Claims also emerged that a full-scale nuclear exchange might kill everyone in the world, meaning there was no meaningful basis for deterrence. Uncertainties and misunderstanding prevailed in science, in international relations, and in the public's mind. The world seemed on a path to a terrible outcome.

Amidst the growing tensions, Professor Zichichi, building on a history of expert meetings devoted to advancing physics, created the series of International Seminars on Planetary Emergencies. I have been fortunate enough to be invited to participate in five of the sessions, each time being very impressed with the very high quality of the participants and the discussion.

My first attendance was at the 3rd Session of the Seminars in 1983 at the height of the Cold War. Despite the many reasons that might have led to governments not allowing their leading scientists to attend, Professor Zichichi assembled leading scientists from the United States, the Soviet Union, Europe, and China. The primary issue being focused on was, as I recall, the emerging issue of the possibility of defensive weapons and how they might affect nuclear deterrence. With Dr. Teller, among others, presenting the case, both scientific and ethical, for moving to nuclear defense and Dr. Velikov, among others, describing the risk that such a step, especially if done unilaterally, would present, bridging the gap with a statement focused on the importance of acting to ensure peace was possible because of both the personal effectiveness of Professor Zichichi and the scientific effectiveness of the opportunity to clarify and exchange views and to build understanding.

My participation, as a climate change scientist from the Lawrence Livermore National Laboratory, was to present a scientific analysis of the not yet published paper that introduced the concept of "nuclear winter," namely that a major nuclear exchange would cause the world to freeze and thus cause the death of virtually all of the world's population that was not killed by the explosions themselves. The initial calculations by Dr. Richard Turco, Prof. Carl Sagan, and others had used a model that represented vertical radiative processes quite well, but did not adequately represent global geography and latitudinal and longitudinal variations. Our model results made clear that the lofting of smoke to above the greenhouse gases would indeed lead to significant surface cooling, but in our model, unlike in the Turco et al. model, ocean heat capacity limited the cooling so that the world did not freeze. In a separate paper, Livermore scientist Dr. Fred Luther presented results indicating the significant destruction of the stratospheric ozone layer that would be caused by the nitrogen oxides injected by megaton size weapons. While these were generally new ideas for nuclear community and the need for further research was clear, the discussion of these early results suggested that the likely global environmental consequences of nuclear war might well tend to deter third party provocation of the superpowers, but that the direct destruction of the combatant nations remained for them the primary deterrent.

My second year of participation was in 1984 at the 4th Session, where the issues were largely the same, but the tensions were higher, leading to the attendees of the Seminar being protected by the Italian Army. The nuclear winter issue received even more attention. My talk and that of others from Livermore concerned new simulations of the climatic effects of smoke that might be generated by a full-scale nuclear war. Results of Soviet scientist Dr. Vladimir Aleksandrov also gave a very significant cooling, although the ocean treatment differed. Again, as for the previous year, the discussions on nuclear winter and other subjects proved very productive, helping to build understanding and clarify misconceptions that often seemed to be generated by imprecise press coverage.

My third year of participation was in 1989 at the 9th Session. By this time, as the Cold War was easing, the primary focus of the Seminars started shifting to a more encompassing discussion of planetary emergencies. My talk was on the scientific basis for human-induced climate change as a consequence of combustion of fossil fuels. While research on climate change had been building over the preceding two decades, it had only just begun to receive the attention of the international policy community. The first assessment of the Intergovernmental Panel on Climate Change was still in preparation and the detection and attribution of human-induced climate change lay even further in the future. As a result, there was both considerable interest and considerable questioning of the scientific basis for the findings and projections of very significant climate change ahead.

In 1993, I again participated, this time because of increasing interest, and even concern, about the potential threat to the Earth's environment and society of an asteroid impact. Our model simulations indicated that substantial climatic effects could result from such an impact, and a paper by Dr. Paul Crutzen reported on the likely substantial damage to the ozone layer. At this session, the set of planetary emergencies was continuing to broaden, bringing together scientists from a widening range of disciplines and so generating very interesting and productive discussions.

My most recent participation was in the 38th Session, held in 2007. The range of planetary emergencies was now, for me, much broader, ranging from brain disease to the energy climate coupling. My talk at this Session focused on the possibility that humans could take actions that would counterbalance human-induced global warming—basically using "geoengineering" to take responsibility for the global climate. While the notion of geoengineering goes back several decades, the accelerating pace of climate change is finally starting to receive broader attention, even though the amount of research that has been done remains quite limited. The result of the presentations on this issue and the discussion led to a recommendation for a coordinated research program in this area that was, following the summer session in Erice, presented in a summary session at the Vatican Academy of Sciences in December.

Across my participation in five of the Sessions (and also a 1986 NATO workshop on climate modeling held at Erice), Professor Zichichi and his excellent staff have been both the implementer and the glue for the efforts, able to bring together experts from around the world and from an increasingly diverse set of fields to form teams of experts around key topics, to provoke very interesting discussion through insightful questions, and to arrange opportunities for the attendees to get to know each other while dining and socializing together (my favorite spot being El Cyclope, overlooking the Mediterranean segment from *The Odyssey*). Through all of these efforts, Professor Zichichi and the International Seminars have created, both in form and in substance, the World Federation of Scientists, a collection of experts around the world who share a common vision for a better and peaceful world (as summarized in the Erice Statement), and who know each other well enough to collaborate through the WFS and on their own initiative in very productive ways.

Bruno Maraviglia

When the Enrico Fermi Centre Council was convened for the first time in December 2000, by the then Minister of the University and Scientific Research, Senator Ortensio Zecchino, I had never met the President of the Council, Prof. Antonino Zichichi (Nino), in person. Of course, I had heard many stories about him reported by colleagues that represented a wide panorama, lacking completely any substantiated base or first hand knowledge. At the same time several of Nino's crucial achievements, such as the Gran Sasso Laboratory, the growth of INFN, the Ettore Majorana Foundation and Centre for Scientific Culture in Erice plus the scientific discoveries, such as the antideuteron, had already made clear to me that Nino was a highly respected scientist who elicited very strong personal feelings.

The activity of the Enrico Fermi Centre Council began without a proper place for its meetings. In fact the first meetings were organized at the CNR (Consiglio Nazionale delle Ricerche) headquarters. Under the guidance of Nino, the Council – knowing that the availability of the historical building of via Panisperna would require some years – converged towards devoting most of its energies and economic resources to some research lines, thus giving birth to the "Centro Studi e Ricerche" in accordance with the founding law. The fundamental subject of the research lines had to be interdisciplinary Physics, resurrecting the idea of Enrico Fermi (1937) of founding an Institute for the application of the new discoveries of Physics.

The enthusiasm of Nino for reviving the historical building of the Enrico Fermi Centre, together with my early role in first officially proposing in a letter to the Interior Minister the transformation of the Panisperna building into a Scientific Institution (1995) – an action which I carried on by urging with the "Comitato Panisperna" the writing and approval of the founding law (1999) – were all reasons which determined a perfect convergence of ideas between Nino and myself, all blended with a growing friendship.

Nino's strong support of the research line "Magnetic Resonance and Brain Function", which I had gradually developed from 1996 on, gave it a crucial impulse with the achievement of a rapidly growing number of publications, together with the formation of a group of very talented researchers. His support was further extended by allowing to me the founding of the Erice permanent school for "Magnetic Resonance and Brain Function," which this year (2009) will hold its seventh meeting. This Erice workshop is now regarded as the most advanced and stimulating event in this field throughout the world. In a way, the direct consequence of my having met Nino was a substantial expansion of all my research activities. Of course my case is not the only one; analogous enthusiastic support was given to other research lines in different areas of interdisciplinary Physics, as it can be verified by the massive contents of the Triennial Research Plan (Piano Triennale di Ricerca) of the Enrico Fermi Centre.

Every Erice Workshop on Magnetic Resonance and Brain Function is a step forward on the path to a full comprehension of the functional organization of the human brain. Any fundamental advance for brain investigation lies in discovering new physical means and consequently in the invention of new original modalities, which allow for observing the physiological and the metabolic processes occurring during brain activity.

Such processes have a time scale, ranging from milliseconds to several seconds, while some of the most successful modalities like Functional Magnetic Resonance are rather slow. It is thus more and more urgent on one hand to find new and efficient magnetic resonance techniques, and on the other hand to try to combine them with the other existing modalities to gather information through the best features of each of them. The Erice Workshop operates also with the target of integrating Magnetic Resonance methods with Electro-Encephalography (EEG) and Magneto-Encephalography (MEG), which have a poor capability of imaging but a higher time resolution with respect to Magnetic Resonance. The confrontation among the speakers has generated important results due widely to the young researchers of the Enrico Fermi Centre. These researchers obtained fundamental results in frontier fields like the imaging of the activated areas of the spinal cord (quite difficult to detect, due on the one hand to the mobility of the spinal cord related to heart beating, and on the other hand to the strong magnetic field gradients associated with the susceptibility variation between the spinal bone and the cerebro-spinal fluid). They reached other important results in the experimental and theoretical investigation of the neuron energetics in the activated brain and, more recently, in the unknown world of brain function in the "resting state", in other words in the space and time connectivity of the brain while it is not stimulated from outside. This last research line opens the new frontier of the way the brain works "by itself", namely without the direct stimulation normally applied in Functional Magnetic Resonance, in EEG and MEG.

My long experience in research ranging from fundamental physics (superfluidity, quantum solids, etc.) to interdisciplinary physics (biological membranes, Magnetic Resonance Imaging methods and systems, Human Brain Function, etc.) allows me to conclude that important achievements can be reached by scientists whose profile includes some fundamental conditions: personal scientific capability is of course the first one; the second is the ability to coordinate and stimulate the research of younger scientists, especially in experimental research; and the last condition - crucial but difficult to satisfy - is encountering other scientists who enjoy their research but at the same time are very interested in advancement of other important fields and supporting by providing them with trust and action. Nino Zichichi is one of these rare scientists. I have personally verified how interested he is in any new important frontier which is brought to discussion; he immediately identifies the essential points of the problem, with an enthusiasm which confirms his youth and great thirst for knowledge; soon afterwards, he starts wondering what could be done to advance this particular line of research.

Sergio Martellucci

Erice is a wonderful city. In addition, I think very highly of the Ettore Majorana Foundation, the Schools and their Courses, the World Federation of Scientists, the International Seminars, the Erice Prize etc., launched by Professor Zichichi. My testimonial is a personal history in part, but also I think a very important illustration of how new science and new technology develop at the service of the great political challenges of our time and our planet. It's simply one of many examples, but one I happen to know personally, so it the the one I have chosen to recall.

You will see how important I believe it is for scientists to interact, to bring people together, to talk with each other, to exchange ideas, and also for different fields to interact, so that one can gather ideas from another field. This is a very important part of the progress of science and the progress of scientists. This is a rule of thumb, a golden rule, at the basis of the success of the various initiatives of Professor Zichichi.

The first time I met Professor Zichichi at the Institute of Physics of the University of Rome, more than 50 years ago, he was working in the cosmic rays physics and I was doing numerical calculations to collect data by the experimental measurements obtained with the use of a Wilson chamber allocated at the Plateau Rosa Laboratory in the Alps. Fascinated by his personality, I studied Physics at the University, and since then he became my guide during my youth and also later on.

In 1957 I joined the selected group of young and enthusiastic researchers who launched in Italy the controlled fusion research programmes, immediately after the *International Conferences on the Peaceful Use of Nuclear Energy* held in Geneva in the late fifties at which were declassified the results of the experiments carried out at Culham in the UK on the magnetic confinement of thermonuclear plasma using the ZETA machine (*Z-pinch experiment*). I was at the time studying for my physics degree thesis, under the tutors Professors Ugo Ascoli-Bartoli and Franco Rasetti who was at the time in Italy on sabbatical leave from the John Hopkins University, Baltimore, Maryland, USA. Professor Franco Rasetti was the "High Priest", "Cardinale" in Italian, of the group of young physicists – known as "the boys of Via Panisperna" – who worked with Enrico Fermi at the old and prestigious Institute of Physics of the University of Rome in via Panisperna where Fermi carried out his first experiments. Professor Rasetti passed away eight years ago (December 5, 2001 at the age of 100).

Because of their in-depth knowledge of optics, acquired previously in Italy and also at German optical institutes, they immediately concentrated on studying optical and spectroscopical methods of measuring the characteristic parameters of a plasma (e.g. density and temperature). It was in 1959 that the electron density of an RF plasma was measured, using for the first time in the world optical interferometry, by U. Ascoli-Bartoli and F. Rasetti (and I was mentioned in the "acknowledgements"). I recall the time when Rasetti took us in a round to the basement of the Institute of Physics in Rome looking for the Jamin interferometer (which was subsequently used in the experiment), that he knew had been delivered to Italy by Austria after the First World War as war reparation. Also then when he looked in the library of the Institute through all the works by Enrico Fermi on interstellar plasma to find the formula for refractivity of a gas of free electrons. In 1960, with the invention of the laser, which Rasetti described as "the most important discovery of this century" they finally knew that they had at last a light source, which would enable to overcome the intense luminosity of plasma and thus measure its density and temperature by Thomson scattering of laser light. From the 1970s onwards, the laser scattering apparatus developed by our group in Frascati, for measuring the temperature of electrons, was also used on the Tokamak Alcator machine at MIT in Boston, USA, and on the Tokamak FT machine in Frascati, Italy, and rapidly became one of the most widely-used methods of plasma diagnosis throughout the world. During the same period, still at the height of the Cold War, the English Culham diagnostic team headed by Nicol J. Peacock, thanks to a special Euratom-USSR Academy of Sciences cooperation agreement, went to what was then the Soviet Union's Kurchatov Institute to measure by "laser scattering" the Tokamak T-3 machine's plasma parameters in collaboration with the Kurchatov diagnostic team.

In 1963, when Nicolay Basov (who in 1964 was awarded the Nobel Prize for Physics) reported, at the *International Conference on Quantum Electronics* in Paris, on the possibility of initiating a new line of fusion research using lasers for the inertial confinement (i.e, the inertial approach to thermonuclear fusion), in Frascati we began to study the resulting phenomenon when a target pellet of deuterium (D_2), solidified at the temperature of 4K, was heated by a laser pulse to thermonuclear temperature. Our typical humour motivated us to call the experiment "Hot Ice", a nickname given at the time to Anita Eckberg, the gorgeous Swedish movie star who played the leading role in the Federico Fellini movie *La Dolce Vita.* She was the cold North-European woman who was "heated" by the Latin lover, Marcello Mastroianni. The Hot Ice experiment, upon request of EURATOM, was closed in 1970, after getting thermonuclear regime and generating neutrons, as the inertial approach was still classified in Italy at that time.

Let us go back to how I interacted with Professor Zichichi, already very famous in those years for his scientific achievements at CERN, Geneva, in particular the discovery of nuclear antimatter. In 1963, the first year of the Erice Subnuclear Physics School, he opened the Ettore Majorana Erice Centre, with today more than one hundred and twenty Schools. Shortly after the discovery of the laser, Professor Zichichi, opened the International School of Quantum Electronics, running Courses since then on the physics and the technology of lasers and their applications under the joint direction of Dr. Arthur Noble Chester, Hughes Electronics Corp & HRL Laboratories (Retired), Malibu, California, USA, and myself.

Among the 46 ISQE courses I directed, the 2004 Course marks a milestone in the history of the laser as well as a milestone in the International School of Quantum Electronics. 2004 marks the 40th anniversary of the award of the 1964 Nobel Prize in Physics "for fundamental work in the field of quantum electronics, which has led to the construction of oscillators and amplifiers based on the maser-laser principle". This award, given to Professors Charles H. Townes, Nicolay G. Basov and Alexander M. Prokhorov, acknowledged the discovery and early development of masers and lasers, which have so richly benefited both scientific research and practical applications. 2004 also commemorates the 40th Course of ISQE, which has presented workshops and short courses on almost every aspect of lasers from 1971 through the present 2009 Course. Because of this fortuitous coincidence, it was decided early in the planning of the 2004 course to set aside time in the program for a ceremony to commemorate the "birthday" of the laser, with the participation if possible of those who contributed to its initial discovery and development. Unfortunately, Professors Basov and Prokhorov are no longer living; however, Professor Andrei Varlamov, on behalf of the Russian Academy of Sciences, agreed to attend the ceremony and give his recollections of these gentlemen and their early years of laser research.

Profesor Townes, who is presently with the University of California, Berkeley, California kindly consented to not only attend the ceremony, but also to give the keynote address. His remarks captured the excitement and energy of the early days of maser and laser research, and also touched on the breadth of applications today. He also pointed out that 2004 was also the 50th year of the operation of the first maser, making it a threefold anniversary.

Professor Townes is one of the scientific world leaders who came to Erice after he was awarded the Nobel Prize in Physics. The 1985 ISQE Workshop on "Progress in Microemulsions" and 1999 Course on "Bose-Einstein Condensates and Atom Lasers" listed among the participants Prof. Pierre Gilles de Gennes (France) and Wolfgang Ketterle (Germany), two of the scientific world leaders who were awarded the Nobel Prize in Physics after their participation in the Ettore Majorana Centre activities in Erice.

After so many years, I felt in Erice as at home, entering into the Spirit of Erice by applying the "culture of love" of the Erice Statement, as well as the respect of the "human dignity" of the Farnesina Statement. Since the very beginning, I served in the Voluntary Scientific Service in Erice, Geneva and abroad, by holding different positions in the various initiatives' promoted by Prof. Zichichi: the World Federation of Scientists, the World Laboratory, the Erice Seminars on Nuclear Wars, the Planetary Emergencies and their Pilot Projects, the E. Majorana Erice Prize Science for Peace.

Besides the above, I was working on my research in the fields of plasma physics and of laser physics and technology, and with my academic career. Winner of a national competition I left the Frascati Research Centre and was appointed Professor of Physics at the University of Naples and, later on, at the "Tor Vergata" Rome University, immediately after its foundation. From 1981 to 1988 President (elect) of the Committee of Physical Sciences of the National Research Council (CNR); from 1987 to 2000 executive officer of the National Institute of Geophysics (ING); and, from 2004 to now, chairman of the National Agency for New Technologies, Energy and Environment (ENEA). I have always succeeded in implementing the cooperation with these institutions and the Erice initiatives with mutual benefits.

I also directed the HEFEST Project, one of the World Lab Pilot Projects, of the "Earthquakes" Emergency, one of the 6 emergencies of the "Global Monitoring of the Planet" Basic Emergency # VII. Furthermore, I participated in the Permanent Monitoring Panel (PMP) on Pollution, coordinated by Richard C. Ragaini and (now) Lorne Everett. In 1999 we organized in Erice the ISQE Workshop on Black Sea Pollution. Besides the above, I supported the actions of Professor Donato Palumbo in the PMP on Energy and in the Direction of the Erice Int'l School of Fusion Reactors Technology (1st Director Bruno Brunelli), by taking advantage of my position at ENEA, the Italian research institution leader in the field of thermonuclear fusion research.

I cannot close this testimonial of mine without mentioning the privilege I have had over so many years in my personal encounters with some of the leading Erice figures. Selecting just a few, on the basis of my personal scientific interests, among so many outstanding people, it is a pleasure to recall, in alphabetic order: Robert Clark, Tsung Dao Lee, Piotr Leonidovich Kapitza, the aforesaid Donato Palumbo, Maurice Shapiro, Edward Teller, Yevgeniy Velikhov, and the last but absolutely not the least, Antonino Zichichi.

I shall conclude with just a few summary observations. Firstly, scientific work should not be supported simply for the technological contributions which can be envisioned, but rather as an exploration which has the possibility of making completely unexpected contribution. We don't know what applied fields might be developed from a given research project. Here again, if you consider microwave spectroscopy of molecules, why would that have been produced by the lasers intense light? That was quite unpredictable, and is not atypical for basic science. If you explore new ideas, you develop new thoughts and new results, and out of these come applications.

Secondly, interdisciplinary interaction can be very fruitful in generating innovative ideas and discoveries. I might say interdisciplinary interaction also reminds me of Erice, this very important place which brings people together, and I think we have to be very thankful for what Professor Zichichi is doing in running the Centre, as well as in organizing it, in bringing people together in different fields, getting scientists from many different countries together. You can see from my illustrations how important this is.

Thirdly, we must be open to viewpoints, ideas and experiments which do not fit accepted scientific thinking and trends, perhaps, we even need to encourage challenges to standard paradigms. In particular, I think that we've got to be very careful overusing committees, which we frequently do. We appoint a committee of important scientists to decide whether certain kinds of science should be pursued, or certain kinds of experiments should be supported. We have these committees, generally composed of senior scientists, and they're very knowledgeable, excellent people, but maybe somewhat fixed in their ideas. We have to listen to them, but I think we've got to be careful not to let senior people completely control the plans for research – we've got to try new and different things.

Finally, there are discussions and interactions between scientists, within the community of scientists, and applied scientists. These interactions are very critical in the development of a new field, and of course that's what allowed lasers to grow so rapidly. The development of any field needs a community – a community of scientists, talking, interacting, each adding their own ideas to it. It's this exciting and productive interaction that we can be thankful for today to Professor Zichichi.

Léon Michaud

The origins of the World Laboratory can be found in the activities of the Erice International Centre for Scientific Culture established in 1963 and in its postgraduate schools, the World Federation of Scientists' activities dating back to 1973, the International Committee Science for Peace created in 1983 and the International Erice Seminars, now at their forty second edition.

In the seventies and eighties of the last century, the scientific community and the world leaders were confronted with the dramatic emergency of growing desertification and its associated biblical calamity, famine, striking the African continent and in particular the Sahelian area, affecting millions of people.

The Erice community led by Professor Zichichi accepted the challenge and through the International Committee Science for Peace, with the encouragement of Italian President Sandro Pertini, developed a programme to combat desertification and famine. The role of science in combating hunger in the world was the subject of a meeting in the Quirinal on 5 March 1984 with President Pertini and Professor G.O.P. Obasi, an eminent African scientist from MIT. On this occasion Professor Obasi mentioned that he had been recently elected Secretary-General of the World Meteorological Organization despite the opposition of many Western countries because of his origin, but that he had had the strong support of Professor Zichichi who had met him when he was lecturing in the Erice International School of Meteorology of the Mediterranean. President Pertini stood up and embracing Obasi said "This is the best compliment for an Italian scientist who had similar problems for his election of President of the European Physical Society".

The programme was further developed by the Erice community in the framework of the International Schools, in particular that of Meteorology. It was officially launched in Geneva on 18 June 1985 under the acronym of Agrhymet, agro meteorology and operational hydrology in the combat against drought and desertification in support of food production and self sufficiency. The programme involved the provision and use of high technology equipment, remote sensing, transfer of new technologies in agriculture, improving weather forecasts and the provision of basic training in these fields. It has been very successful and the Centre in Niamey is still operational for the benefit of all African countries.

In July 1985, at the opening in Trieste of the meeting for North-South and South-South Cooperation, Professor Zichichi had several discussions with the Nobel Prize winner Abdus Salam, President of the Third World Academy of Science, and the newly appointed Secretary-General of the United Nations, Javier Pérez de Cuéllar, on the strong need of implementing the idea of laboratories with open walls launched in 1950 by Niels Bohr in a famous open letter to the United Nations, and reaffirmed in the 1982 Erice Manifesto. Further discussions took place in the plane from Trieste to Rome and in high level meetings in Rome. Javier Perez de Cuéllar, late in the evening on the eve of his departure from Rome, expressed his regret of not having been able to visit the Vatican and the Vatican Museums. He was amazed when Professor Zichichi arranged immediately for an exceptional night visit to the Vatican.

At the Fifth session of the Seminar on Nuclear War, in August 1985, the idea of an international laboratory without walls, without secrets and without frontiers, was further developed by Professor Zichichi and other scientists, in particular Edward Teller and his collaborator Lowell Wood, Eugene Wigner and John Eccles. Main supporters of the concept were the Italian Foreign Minister Giulio Andreotti, Ambassadors Qian Jiadong of the People's Republic of China and Ambassador Henning Wegener of the Federal Republic of Germany, and chief advisor to Chancellor Kohl. For reasons not yet completely clear, the Soviet delegation was not in attendance at the seminar despite the assurances given by the newly appointed Foreign Minister Eduard Shevardnadze. However, Professor Zichichi met Academician Velikhov in Geneva in November 1985 on the occasion of the Gorbachev-Reagan summit.

At the invitation of the Academy of Sciences, in January 1986, Zichichi went to Moscow where he was given the assurance of the willingness to continue the Erice dialogue between scientists and to substantially participate in the activities of the planned World Laboratory. In fact, Soviet scientists actively participated in the August meeting and in the International Conference on the Peaceful and Safe Use of Nuclear Energy, organised in Rome in November 1986 in the framework of the activities of the Erice International Seminars on Nuclear War. An important development of the new East-West scientific relations in the Gorbachev era was the release, on 14 December 1986, of the eminent scientist Andrei Dmitrievich Sakharov from his exile in Gorki (now reverted to Nijny Novgorod).

The news was communicated by telephone, by Academician Evgeny P. Velikhov and Deputy Foreign Minister Vladimir F. Petrovsky, to Professor Zichichi, the leader of the untiring campaign conducted by the World Federation of Scientists in support of Sakharov. In discussions with Velikhov in Moscow and Rome, and in Erice with Petrovsky, the special envoy of Gorbachev to the meeting of experts on strategic stability, Zichichi emphasised that the Sakharov issue was a major obstacle to scientific East-West relations. A confidential and lengthy discussion with Petrovsky took place on the night of 26 April 1986, while walking in the narrow streets of Erice, on the day of the Chernobyl nuclear accident. On the occasion of his honorary degree from the University of Bologna in February 1989, Sakharov expressed his gratitude to Zichichi and his willingness to participate in the activities of the World Laboratory in the field of safe nuclear reactors. Unfortunately Sakharov died on 14 December 1989. Exactly three years after his liberation from Gorki.

A first group of World Laboratory projects was launched in Erice in February 1986, following the proposals of the Fifth session of the International Seminar. However, the Laboratory was legally established at CERN in Geneva, on 12 July 1986, as the International Centre for Scientific Culture – World Laboratory, in short Worldlab. The Erice Statement was the basic principle underlying its activities for East-West and North-South cooperation in science, without secrecy and frontiers. The Constituent Instrument set the following objectives:

- To promote East-West and North-South scientific and technical cooperation for the purpose of conducting scientific research without secrecy and without frontiers;

- To favour the free circulation of scientific information, scientists and researchers, in particular by means of fellowships and visiting scientists' programmes;

- To carry out projects in the scientific, technical, medical, meteorological, environmental and other fields, in particular for the benefit of the developing countries, associating thereto researchers from other Organizations and Institutions without any discrimination, whether geographical, political, ideological, religious or racial.

Many scientific institutions and research centres in the world were represented in the Constituent Assembly, and signed the Constituent Instrument. These institutions included the Academia Sinica, the African Academy of Sciences, the Third World Academy of Sciences, the International Centre for Scientific Culture Ettore Majorana, the International Centre for Theoretical Physics, the International Centre of Insect Physiology and Ecology, the Foundation Jean Monnet for Europe, the Foundation Galileo Galilei, the Institute of Fundamental Studies of Sri Lanka, the Permanent Interstate Committee for Drought Control in the Sahel (CILSS) grouping nine African countries. Eminent scientists such as Nobel Prize winners Tsung Dao Lee, Samuel C.C. Ting, Robert Hofstadter and Professor Donato Palumbo, also signed the Constituent Instrument. From Sri Lanka Professor Cyril Ponnaperuma sent an enthusiastic message of support.

The surprise of this memorable day was the invitation from Foreign Minister Andreotti to all participants in the Constituent Assembly to have a working lunch at Villa Madama in Rome with a special return flight and quick crossing of the heavy Saturday traffic in Rome.

The Scientific Committee of the World Laboratory approved an ambitious programme for development cooperation activities, for the period 1986-1988, with a fellowship scheme involving the granting of one thousand international fellowships and fourteen large development cooperation programmes in fourteen key emergency sectors. Certain projects were concerned with advanced scientific and technological fields and the protection of the environment and climate. In particular, the establishment in Beijing of a Centre of Advanced Science and Technology, the improvement of flood forecasting and control of the Huang He (Yellow River), application in China of coal slurrying technologies, applications of seismological research in developing countries with high seismic risk, optimum use of environmental resources to develop agriculture and animal husbandry in the Sahel, applied research on drought and desertification, protection of coastal marine environment.

Other projects concerned key sectors of the medical sciences for the Third World such as the applications in China of advanced biotechnologies, the study of AIDS and related human oncogenic viruses in Africa and establishment of local units of epidemiological control, epidemiological studies and the training of staff for the prevention and treatment of heart diseases in African countries, survey of elderly populations and training of social health assistants, radio-diagnostic assistance and training of technical and medical staff in China and Africa, prevention and treatment of deafness in developing countries. Project directors were eminent scientists in the medical field such as Robert Gallo, Luc Montagnier, Guy de Thé, Giovanni Motta, just to name a few.

The then Secretary-General of the United Nations, Javier Pérez de Cuéllar, confirmed his support to the World Laboratory as an international scientific organisation in consultative status with the Economic and Social Council. The headquarters of the World Laboratory were established in the historical Palais de Rumine in Lausanne with a permanent secretariat, headed by Claude Manoli.

The implementation of projects started quickly. On 17 October 1986 the Centre of Advanced Science and Technology was opened in Beijing by Professor Tsung Dao Lee. On this occasion, Chairman Deng Xiao Ping received the scientists of the World Laboratory and the event was reported with great evidence on the first page of the People Daily. Other projects were also very successful, such as the flood forecasting and management for the Yellow River, an ancestral emergency of China, the coal slurrying project - anticipating the present research on clean carbon technology. To be mentioned also was the combat against drought and desertification in Sinkiang, carried out by outstanding scientists such as Professors Duzhen. Ye, Qing-cun Zeng and Zhu Zhenda, who made pioneering studies on the natural variability of global climate and its anthropogenic components. In the combat against desertification, successful experiments were conducted in the Shapotu desert area in the use of sand-fixing plants such as Artemisia Ordosica and Caragana Korshinskij. Also of note in this connection is the volunteering of all of the scientific staff involved and project directors, in particular Professor Robert Clark of the University of Arizona and the late Professor Giovanni Novelli, scientist of the clean carbon technology, also known for his Carbon 14 studies for the dating of the Holy Shroud.

The opening of the Moscow branch, on 27 February 1987, in the presence of Foreign Ministers Shevardnadze and Andreotti, was a milestone event in the development of the World Laboratory. The President of the Academy of Sciences, Gury Marchuk, stated in his introductory speech that the opening of the World Laboratory in Moscow was a significant event in East-West relations and marked a new stage in the development of international cooperation in science, in line with the proposal supported by Mikhail Gorbachev for the creation of laboratories without walls and without secrecy and frontiers. The projects approved for implementation focused on seismology, electromagnetic sounding, research on earthquake predictions and the development of broad band digital seismographs. The study of solar neutrinos under the Gran Sasso was also included in the scientific programme as well as research on controlled nuclear fusion under international aegis. The headquarters of the Moscow branch were temporarily accommodated in the building of the Institute of System Studies, directed by late Academician Djerman M. Gvishiani.

The Seventh session of the International Seminars in August 1987 was entirely devoted to the great projects of the World Laboratory for scientific cooperation East-West, North-South. Messages of support were received from the world leaders and from His Holiness John Paul II. That of President Reagan was conveyed by Edward Teller. A four-meter parabolic antenna had to be installed on the roof of the San Rocco Institute to receive, through the Horizon satellite, the message of Gorbachev read on Soviet television. Special transportation by an exceptionally large truck had to be organised, from the manufacturer's premises in Northern Italy all the way to Erice during the Italian mid-August holidays.

A Liaison Office with the World Laboratory was established in 1987 in the Secretariat of the World Meteorological Organization, with the support of the Secretary-General, G.O.P. Obasi and the President Zhou Jingmeng, Administrator of the China's Meteorological Organization. Project proposals were developed for base line carbon dioxide and pollution monitoring, environment protection in the Urals and in the Caspian and Aral Seas area in collaboration with Professor Yury A. Izrael, Chairman of the State Committee on Hydrometeorology and Environment Control, Professor Umirzak Sultangazin, President of the Academy of Sciences of Kazakhstan, and Victor E. Chub, Minister in charge of Meteorology and Environment in Uzbekistan An oceanographic expedition along the Northern Sea Route was also planned by Polar explorer Arthur N. Chilingarov to study the effects of global warming in the Arctic regions.

The World Laboratory branch of Central Europe was opened in Trieste on 12 June 1989 by Italian Foreign Minister Giulio Andreotti and Nobel Laureate Abdus Salam, Director of the International Centre for Theoretical Physics. A new series of projects were launched for planetary monitoring of meteorological and climate parameters, climate research, and an ambitious programme of controlled nuclear fusion research.

The projects of the World Laboratory were included in international agreements such as the Declaration of Intent of 30 November 1989, and the Twenty-Year Friendship Treaty between Italy and USSR, signed in Rome on 18 November 1990. The signature of this Treaty has been a major event in the history of the World Laboratory since a specific article provided for full cooperation in science and technology through the World Laboratory. Very impressive was the solemn ceremony of the signing that was transmitted via Mondiovision. The President of the World Laboratory was near the Italian President Francesco Cossiga, the Prime Minister Giulio Andreotti, and the Presidents of the Senate and of the Parliament. Mikhail Gorbachev, President of the Soviet Union, was escorted by Foreign Minister Eduard Shevardnadze and Marshall Dmitri Timofeievich Yasov in full uniform with both sides of the breast covered by medals and war decorations. This occasion was a major event in East-West relations in which the Erice International Seminars and the World Laboratory played an important role. Professor Zichichi had follow-up discussions with President Gorbachev in the Quirinal, Rome, and later on in the Kremlin, Moscow on 13 December 1990.

Some of the projects for East-West cooperation, such as electro magnetic soundings of the earth's crust in the Kola Peninsula, were successfully implemented. A World Laboratory Centre was established in the historical town of Pereslavl Zalevsky to monitor radioactivity and the effects of global change on the ecosystem and biodiversity. It was planned to restore a monastery to host the Russian headquarters of the World Laboratory. The short-lived coup-d'état of 19 August 1991 which caught by surprise the Soviet Delegation in Erice, and the collapse of the Soviet Union in December 1991, prevented the completion of the other projects thus marking a standstill in scientific cooperation.

Shortly after the implosion of the Soviet Union, the East-West cooperation started again in a decentralised manner with the Newly Independent republics. A meeting of scientists of these new republics took place in Erice from 12 to 15 March 1992. Scientific institutions of Bulgaria, Czechoslovakia and Romania also participated. Many projects were developed by the Moscow branch of the Russian Federation and presented by Academician Evgeny Velikhov. Very active branches of the World Laboratory were opened in Lithuania, Estonia, Ukraine and Georgia.

Particularly important for the scientific community was the research work on the nuclear accident of Chernobyl of the 26 April 1986 and on nuclear safety carried out by Professors Victor G. Baryakhtar, Valery P. Kukhar and Genadiy I. Palshin of the Ukrainian branch in cooperation with Professor Yury Antonovich Izrael of the Russian Academy of Sciences. A team of the World Laboratory visited the Chernobyl nuclear plant in November 1992.

The Lithuanian Branch developed notable programmes in the field of read-out electronics, safe and clean energy sources and ecological regional developments, by Academicians Juras Požela, Zenonas Rudzikas and Leonardas Kairiukstis. The Estonian branch, directed by the late Academician Karl Rebane, was very active in spectrophotometry and photochemistry. Many other projects in pioneering sectors of science, technology and medicine were subsequently approved and implemented despite the strict budgetary limitations.

A special session of the Seminars on Planetary Emergencies took place in Erice on 8 May 1993 to celebrate the thirtieth anniversary of the Erice International Centre. Professor Zichichi opened the session by addressing His Holiness Pope John Paul II on the danger of the environmental holocaust and on the achievements of the World Laboratory in the combat against planetary emergencies. On the podium were Professor Luc Montagnier, Nobel Prize winners Tsung Dao Lee, Samuel C.C. Ting, Kai M.B. Siegban, Professor Edward Teller and Academician Yury A. Izrael. In the audience in the P.A.M. Dirac Hall in the San Francesco Institute were the Gotha of science, the directors of the World Lab projects, the Director-General of CERN, Luciano Maiani. The President of the Academy of Sciences of Kazakhstan, Sultangazin, came especially from Almaty to recall to the special session the environmental tragedies of the Caspian and Aral Seas and of the five hundred open-air nuclear tests in the site of Semipalatinsk. His Holiness the Pope was escorted by Monsignor Stanislaw Dziwisz, now Cardinal of Cracow.

It is now time to assess the results of the World Laboratory 23 years after its formal establishment in Geneva. No doubt that the achievements are very positive and leave an impressive and long-lasting legacy to future generations of scientists. About 120 projects in key sectors of science, technology and medicine, were successfully implemented in 50 countries, with the involvement of hundreds of scientists of the World Federation and the active contribution of the 120 postgraduate schools of the Erice Centre. An impressive number of surveys and project proposals were developed by the scientists of the World Federation. The granting of 5000 national fellowships has also helped to fill the gap between North and South.

The climate issue has always been given priority attention by the Erice International Seminars As indicated in a message of Academician Yuri Izrael on the occasion of the twentieth anniversary of the establishment of the Intergovernmental Panel on Climate Change (IPCC), the antagonist role of carbon dioxide and aerosols in climate models has been at the centre of the attention of the Erice scientific community since the early 1980s. The idea of having a panel of scientists assessing the nature and the origin of climate change and ways and means of mitigation and adaptation was born in Erice and supported by eminent scientists, such as Tsung Dao Lee, Edward Teller, Michael C. McCracken, Paul J. Crutzen, Gury I. Marchuk, Georgi S. Golitsyn, Evgeny P. Velikhov, Mikhail Ch. Zalikhanov, Rumen Bojkov, Mikhail Ja. Antonovsky. and Yury Izrael himself. As a result of this proposal, IPCC was formally established in Geneva in June 1988 as a UN body which was awarded the Nobel Prize for Peace in 2007. However the Erice community deserved some recognition for its pioneering role, concluded the message of the former Vice-Chairman of IPCC.

In conclusion, the World Laboratory as an outcome of the Erice International Seminars proved to be a powerful instrument in the fight against the Planetary Emergencies. If this combat is to be continued successfully a revival of the activities of the World Laboratory is needed.

Akira Miyahara

My first encounter with the World Laboratory took place in 1987 at the Hefei Anhui province in China while I was strongly involved in the Chino-Japanese Collaboration programme in the Nuclear Fusion Area. The Cold War status was still continuing, and COCOM did not allow us to bring computers into China although they were purely for academic purposes. The World Laboratory, however, succeeded in providing the Institute of Plasma Physics, Chinese Academy of Sciences, with IBM computers, thanks to the intensive efforts of Professors Donato Palumbo and François Waelbroeck. I think that this achievement of the World Laboratory was one of the best cases of breakthrough during the Cold War period.

I have attended all the Erice Seminars on the Problems of Nuclear War and Planetary Emergencies since 2002. I have, however, had the impression that many of the plenary talks presented at such seminars reflected, more than was necessary, the political ideology of the states to which the speakers belonged. In my opinion, the speakers who already know how the authorities of their countries have reacted to the contents of their talks should do their best to provide the Erice audience with full explanation of such governmental reactions, to secure complete understanding among the participants at Erice Seminars. At the same time, it is necessary that internationally acceptable recommendations should be developed and that such recommendations should be informed to the Directorate of the Erice Seminars in a document as evidence of the discussions about, and its resolutions of, all related problems at Erice plenary meetings. In this sense, the role of PMPs (Permanent Monitoring Panels) is important, not only for preparatory work to organize plenary talks with clear problem definitions, but also to conclude the remarks to the next step.

At the end of World War II the US forces killed large numbers of Japanese citizens by dropping nuclear bombs on Hiroshima and Nagasaki just after it ascertained that the Nazis were no longer able to resist the Allies. Professor Einstein, who had earlier warned that Germany would probably resort to nuclear warfare, was infuriated when he learned of the said brutal act of the US forces, and began formally accusing the US of its act of brutality together with Bertrand Russell and many other scientists of the world. I do not know whether any detailed record of nuclear attack disasters at Hiroshima and Nagasaki is attached to the proceedings of any past Erice Seminars on Nuclear War and Planetary Emergencies. If there is no such document, then it will be necessary to secure at least one testimony based on a physicist's personal experiences of the casualties at these cities because it was the result of the first nuclear war humanity has ever encountered.

Last of all, I should like to pay a sincere tribute to Nino. We only have to read his opening talks he has been giving at Erice annual Seminars, to find his ardent faith in Science. His speech always encourages me to renew my determination to be true to the ERICE SPIRIT. We sincerely hope that he will enjoy good health and maintain his firm faith in science in order to stimulate our spiritual feebleness.

Povl L. Ølgaard

In 1997 and 1998 I had the pleasure to participate in the Erice International Seminars held at the Ettore Majorana International Centre for Scientific Culture in Erice. Here many subjects of great importance for the safe development of the world were presented and discussed. Just to mention a few I recall the discussions on the threat of the weapons of mass destruction, the proliferation of nuclear weapons, the mad cow disease and the effective detection of land mines. Some of these subjects are essential to the peaceful development of the world, some are perhaps not of major global importance, but still of great significance to groups of people on this earth. What was also important for a fruitful discussion was that the participants came from many different countries, both from the major powers and from the smaller countries and both from the West and from the East. Let me just mention a few of the participants: Professor Siegbahn from Sweden, Dr. Velikov from Russia, Dr. Morrison from CERN and of course the man who arranged it all, Professor Nino Zichichi. Many of the participants were, while primarily scientists, also advisers to their governments. Therefore the presentations and the discussions at the seminars did not only increase the insight of the participants, but were also beneficial to wider circles.

The threat to the world due to the risk of an armed East-West confrontation, including the risk of a major nuclear conflict, has greatly decreased after the end of the cold war, and the Erice Seminars have made their small contribution towards this development. Unfortunately the threat of nuclear proliferation and thereby the risk of local nuclear wars has recently greatly increased.

The nuclear cooperation agreement between the US and India, which was signed in October 2008, will almost unavoidably contribute to the undermining of the NPT. The basic purpose of the NPT was to limit the number of countries with nuclear weapons to the five, who had these weapons at the time the NPT came into force in 1970. The US-India agreement and its acceptance by the Nuclear Suppliers Group means the formal acceptance of India as the sixth nuclear weapon power. In addition the build-up of the Indian nuclear arsenal has so far been limited by the limited Indian uranium deposits. Now India can import uranium for its civil nuclear power stations and reserve its own uranium deposits for military purposes.

The US-India agreement has naturally had the effect that Pakistan and Israel have claimed that they should be allowed the same type of agreement as India. So far the US has resisted these demands, but considering the US need for good relations with Pakistan in connection with the on-going conflict in Afghanistan, and the traditional good relations between the US and Israel, it is an open question how long time the US can resist giving in to these demands. So the world may soon have eight formally accepted nuclear weapon states, and that will hardly be the end – except to the NPT.

The next nuclear weapon state could well be North Korea who will undoubtedly also ask to have the same type of agreement and such a development is likely to make it impossible to have Iran give up its uranium enrichment work. On the contrary, it may well encourage Iran to go ahead with the development of nuclear weapons.

If this development materializes it is most unlikely that it will stop at this point. Countries in the Middle East may well feel that they will have to acquire nuclear weapons to deter the threat from the Israeli and future Iranian nuclear weapons. Countries in the Far East may well decide that they will have to counter the threat of the North Korean nuclear weapons by acquiring their own nuclear weapons. And the more countries that acquire nuclear weapons, the higher the probability that these weapons will be used with disastrous results.

The end of the cold war meant that nuclear conflict between the East and the West has become much more unlikely. As a matter of fact, nuclear weapons play today a much smaller role in the East-West relations than they did during the cold war. They have almost been forgotten. Therefore it is a most unfortunate development that the US-India agreement is undermining the NPT and making the acquisition of nuclear weapons by new countries and – even worse – making the use of these weapons more likely.

Such a possible development makes it essential that scientists meet and discuss the future possibilities to ensure development in a peaceful direction. And of course there are other threats to the world,for example, climate changes which can also have disastrous effects on mankind, and which also need to be considered. So there is still a great need for meetings like the Erice Seminars to help to ensure that the world develops in a safe direction

Let me finish by mentioning that the spirit of the Erice Seminars was beneficially influenced by the wonderful town of Erice, this old medieval town with its old buildings and city wall situated on the top of a mountain with the marvellous view over the surrounding country. The same is true for the delicious food which was served to us not the least at the Al Ciclope restaurant. I should also mention the excursion which the participants were offered. At the first seminar we were taken to a site with remains from the roman times of which Sicily have so many, at the second seminar some of us went to Palermo where we visited a beautiful church. These additional features created a very friendly atmosphere, so that even if the participants did not fully agree, the discussions were always held in a objective and friendly manner.

Thus I have every reason to be thankful to my good friend Nino Zichichi, whom I also have known during the many years we were both members of the NATO Science Committee, for arranging these very informative and very pleasant Erice Seminars.

Jef Ongena

The Erice Seminars on Planetary Emergencies organized each year by Prof. Zichichi and his team offer a unique forum for scientists to broaden views in their own field of research by the exchange of ideas with high-level colleagues in complementary domains of science. The uniqueness of the seminars is difficult to summarize but could perhaps be exemplified by the following three main characteristics: (i) the free exchange of ideas, with participants from all over the world (ii) the interdisciplinarity and (iii) the long standing tradition which also made them one of the first where global emergencies could be discussed, on a regular basis. That the seminars also took place during the cold war years clearly shows their importance and impact.

The friendly atmosphere of the Seminars offers in a natural way plenty of possibilities for in depth discussions with high-level scientists. The enjoyable climate of Sicily and the excellent organization contribute largely to create an atmosphere of a big family. The participants therefore have access to a source of information that is difficult to find anywhere else. I always return home with the pleasant feeling of really having learned and understood particular scientific issues that previously looked incomprehensible and difficult.

The Seminars made me aware of the importance of global emergencies that I had not recognized as such. They are the ideal setting for someone eager to learn, with the interdisciplinarity enabling "cross-fertilization" between different fields of science. Being a fusion scientist and member of the Energy PMP, I enormously enjoyed in- depth exchanges with scientists involved in climate research. This is not my field of research but there are close links and the Seminars are therefore were an invaluable source of first hand information on the uncertainties in understanding and modeling global climate change and the real impact of global warming. Where else would I get such an opportunity? The Seminars also introduced me to many other subjects ranging from virus diseases such as AIDS, or new threats like the H1N5 flu virus, brain diseases like e.g. Alzheimers' disease, the danger posed by meteorites, the ecology of the Black Sea (with the fascinating possible link to the Big Flood), the enormous pollution of the oceans by plastic waste and the damage it causes to wildlife, progress in recultivating desert land by very simple but effective means, etc. Each of them (and many more) awakenedrose my interest, which in itself is a very enriching experience.

Encounters and contacts made in Erice are experiences that last a lifetime. I had the privilege to work very closely with Prof. Kai Siegbahn as a member of the Permanent Monitoring Panel on Energy. It was a thrilling experience to meet this famous scientist and Nobel Prize winner, and moreover then to learn from his experience. As a consequence of the frequent encounters I got to know his family and had exchanges with his laboratory. Douglas Morrison is another colleague from the Energy PMP that made a lasting impression. He was extremely well documented on all aspects of energy research, and acted as such as the 'encyclopedia' of our PMP. Discussing with him was a real delight, and I will never forget the evenings with him during dinner. Thanks to the Seminars, I got in contact with the family of the famous scientist Andrei Sacharov. His widow, Elena Bonner, gets now every year an update on what is going on in fusion research. These are just a few examples of the many unexpected contacts the Erice encounters offer.

The impact of the Seminars reaches further than may seem at first sight. First, they represent a sign of hope in a time with so many negative messages for the future. Problems are tackled, possible solutions discussed, new ideas proposed and scientists find here the necessary forum to draw attention, and discuss the emergencies they recognize as important and to make recommendations to their governments. Again, where else is this possible? Second, the general public also benefits from these meetings. E.g. For example, in my case, new insights gained on climate change and world energy issues allow me to reach out to students and general public with much more authority than was otherwise possible.

Andrei Piontkovsky

When Harry Truman was President of the USA, he liked to remind the world about America's monopoly of nuclear weapons. "I need a big stick against these guys," he said in 1945, having in mind the Soviet leaders of the time.
After the Soviet Union acquired atomic weapons and launched the first Sputnik, the Kremlin leaders basked for a while in nuclear euphoria. "We are turning out nuclear missiles as if they were sausages," Khrushchev boasted. His aspiration was to "put a hedgehog in the Americans' pants," and this is just what he did in the autumn of 1962, bringing the world to the brink of catastrophe.

The military and political leaders of both superpowers emerged from the Cuban missile crisis maturer and wiser, and it was then that the famous doctrine of Mutually Assured Destruction (MAD) was born. American presidents and Soviet Party leaders never again threatened each other with nuclear weapons, well understanding that they had no big stick they could use against each other, and that their two countries were chained together by their nuclear potential, their necks in a single noose which would enable them to commit suicide together at any moment they chose.

This clear understanding was a positive force in the Cold War era, deterring the USSR and USA from going head-to-head even in the tensest of political situations.

Reagan's Strategic Defense Initiative altered nothing in the paradigm of Mutually Assured Destruction. The Star Wars program, suggested by the famous physicist Edward Teller, creator of the American hydrogen bomb, was an enormous bluff but this bluff was taken seriously by thirteen extremely elderly people. One of these was Ronald Reagan, President of the United States, who allocated colossal funding to it in 1983; the other twelve were the members of the Soviet Politburo, who lost their cool and started rushing about like headless chickens.

In 1993, an annual Russo-American seminar on nuclear strategy was held in Erice under the auspices of Professor Antonino Zichichi' "World Laboratory". It was chaired by Edward Teller and Yevgeny Velikhov. In relaxed hospitable surroundings of our Piano Room, I asked the eighty-five-year old patriarch of American physics, "How on earth did you manage to persuade Reagan of the feasibility of anti-ballistic missile defense?" "Young man," he replied, "I very much needed money for interesting physics." Those were the words of a true scientist.

The Cold War has long been over. Relations between Russia and the United States, if still not straightforward, have changed radically, but the nuclear arsenals they inherited have remained essentially as they were. Just as forty years ago, we retain the capacity to destroy each other. This situation is set to continue for at least the next ten to fifteen years, judging from the extreme inertia of the plans that both powers have for modernizing their strategic nuclear forces.

There would be no need to labour these obvious facts had not our Commander-in-Chief developed a curious habit in the past year and a half of repeating a particular sentence, full of abstruse but completely meaningless terminology, about "tests of unique new missiles capable of penetrating inter-continental depths at supersonic speeds and of defeating any anti-missile system."

Some bright spark penned this masterpiece for the President, he learned it by heart, and now solemnly repeats it on every conceivable occasion, inserting it, for example, into his replies to "questions from citizens." He seems unaware that there is nothing unique about intercontinental missiles traveling at supersonic speeds on the ballistic segment of their trajectory, or that their ability to confound any particular configuration of anti-missile defense derives not from the characteristics of an individual missile but of the entire network of strategic forces. If he is referring to the anti-missile system which the Americans are trying to develop as a defense against possible terrorist strikes, then it is well known that present-day Russian strategic forces are already capable of penetrating it with ease.

The political illiteracy of this utterance is far more serious than its technical illiteracy. In effect, Russia's Commander-in-Chief is repeating with relish every two or three months, "We can destroy the United States of America." This is a regression to the level of Khrushchev's sausages and hedgehogs.

Brezhnev never came out with any such nonsense. No doubt advisers of a somewhat higher calibre than the present clerks from the St Petersburg Mayoralty explained to him that, correctly expressed, the sentence would run somewhat differently: "We and the United States of America are unquestionably capable of destroying each other." This is nothing to be pleased about. It is nothing to be proud of, period.

Juras Požela

Along with other lucky people, who have had the opportunity to participate in the process started with the Erice Statement drawn up by Paul A. M. Dirac, Piotr Kapitza and Antonino Zichichi, I have been participating in its implementation fora – the Erice International Seminars – for quite a number of years. I have today the opportunity to testify of the truly extraordinary and responsible people, scientists and thinkers of highest rank attending, the significant ideas exchanged and spread, and the great things happening at these Seminars.

In the end of 20th century, confrontation of two political systems, involving new acceleration of military technology race, had become extraordinarily dangerous and insane in the eyes of scientists. And they said their word on both sides of the then existing "Iron Curtain". The political leaders, made aware of the critical situation, came down to allowing the global open scientific discussion of this issue. For me and my colleagues from the then-USSR, this started with the participation in the Global Consequences of Nuclear Winter Seminar in October 1986, organized in Geneva by the European Physical Society. As you may recall, Antonino Zichichi had been the founding President of this Society.

Although no governmental decisions were announced during those Cold-War period meetings, the leaders of the states involved were well informed on the opinion of scientists, and apparently took notice. With Mikhail Gorbachev's policies of perestroika and glasnost, changes came about in the foreign policies of the USSR: the new government intensely promoted the disarmament, the USA government welcomed that, and the process started at last. But, as we know now, this ultimately was no reason for the world scientific community to relax: the Erice Statement proposal on free flow of information, ideas and people helped to further identify the other emerging global threats.

The later Seminars on Planetary Emergencies held in Erice, similar to the pre-détente meetings, have assembled the highest ranking scientists and public figures who had adopted the ideals of the World Federation of Scientists. Nobel Laureates such as George Charpak, Kai M. Siegbahn, Samuel C. C. Ting, Tsung-Dao Lee and Sir John Eccles, and famous scientists such as Edward Teller, an adviser to President Ronald Reagan, professors Yevgeni Velikhov, Robert Clark – to mention just some – were heard during the sessions and informal meetings. Kai M. Siegbahn in fact had become a supporter of our research project on amplifiers for GaAs detectors, carried out at the Lithuanian Branch of World Laboratory.

The International Centre for Scientific Culture – World Laboratory (ICSC–WL), helping to achieve the goals outlined in the Erice Statement and promoting research without secrets and frontiers, established its Lithuanian Branch (LB) in 1992. Back then, after the collapse of the USSR, this was an important incentive to perform internationally significant research in Lithuania. In particular, besides the GaAs project, related to international collaboration "GaAs electronics" led by the World Laboratory President Antonino Zichichi, the WLLB carried out a "Lithuanian Ecological Sustainability" programme. In the latter, more than 200 Lithuanian scientists from not less than 20 institutions cooperated while aiming to implement the World Federation of Scientists ideas of global sustainability on the local scale. Appreciating these activities of the World Laboratory, the Lithuanian Academy of Sciences elected Antonino Zichichi its Foreign Member in 2005.

The International Seminars in Erice have been attended by a number of scientists from former

Soviet Union Republics. We have always enjoyed seeing them and continuing to discuss subjects of common interest, including latest anecdotes. We contributed actively to the seminars and it is now public knowledge that the emergence of global terror acts was predicted at an Erice International Seminar by Karl Rebane, from the Estonian Academy of Sciences. On the grounds of the principle of entropy increase, he had argued, even before September 11, 2001, that the variety in ways of peoples' life should be respected and protected.

One of the brightest encounters not to be omitted from this testimonial has been with His Holiness John Paul II, who participated in two events in Erice and kept close contact with WFS. Here are the words of the Pope: "The enemy, ruining the progress and peoples' lifes today, is named selfishness, lack of love to one's brother, strive for unlimited power in all domains: economics, politics, and production. A firm turn in culture is urgent; in culture, the science is not the opposite of faith, but an ally, helping to raise citizenship, morality, adding much to the growth of well-being – not only of separate individuals, but of all men ond women on Earth, who will meet in brotherhood and solve the problems of global emergency threats."

And one more true story to this. When His Holiness John Paul II had been visiting the house of the Lithuanian Parliament on the occasion of his visit to Lithuania, he was shaking hands with every member with sincerity so characteristic of him. Having recognized me, he smiled. The journalists were puzzled for a long time over the two unintelligible secret words we have uttered then in exchange. I said: "Erice!" The Pope replied: "Zichichi!"

Richard C. Ragaini

My participation in the Erice International Seminars has been one of the most rewarding experiences of my scientific career. Through Professor Antonino Zichichi and the Erice International Seminars I have met hundreds of scientists, including the Nobel Laureates, T.D. Lee and Kai Siegbahn, from all over the world. I have participated in numerous exciting workshops and conferences in Erice, Rome, Lausanne, Gran Sasso National Laboratory and Geneva. The Seminars opened valuable paths of professional and personal discovery for me.

My involvement with the Erice International Seminars dates from 1989, when I was invited by Dr. Edward Teller to the annual WFS International Seminars on Nuclear War to present a talk on "Destruction of Chemical Weapons Using Incineration." The talk was part of a joint US – USSR session on chemical weapons destruction technologies. This session was extremely important in 1989, because it was an example of scientific cooperation on weapon destruction in the midst of the cold war. That year I co-founded the WFS International School of Innovative Technologies For Cleaning The Environment with Nikolai Plate of the USSR Academy of Sciences and William Barletta of LLNL. As the co-director of the School, I organized Erice meetings in 1990, 1992 and 1995, which are discussed below.

The October 1990 meeting sponsored by my WFS International School of Innovative Technologies For Cleaning The Environment was the workshop, "Innovative Cleanup of Contaminated Soils and Groundwater." This was the first WFS meeting to discuss technologies, which could successfully clean up contaminated soil and groundwater. At that time in Italy it was widely believed that it was impossible to clean up groundwater after it had become contaminated, so there was heightened interest in our meeting. Professor Zichichi personally participated in the meeting, which was significant, because it was his first environmental restoration workshop. Reporters from the Italian TV program Green Line attended the meeting and filmed interviews with Professor Zichichi, myself and Igor Zetsker of the USSR Academy of Sciences, the leader of the USSR delegation. Also interviewed were Mordecai Margaritz of Israel, Lorne Everett of the USA, and Fred Hoffman of LLNL. These interviews were later televised in Italy by the Green Line program.

The April 1992 meeting sponsored by my WFS International School of Innovative Technologies For Cleaning The Environment was the conference, "Innovative Technologies for Cleaning the Environment: Air, Water and Soil," which was keynoted by Dixie Lee Ray, former head of the US Atomic Energy Commission and Leo Duffy, head of the US Department of Energy Environmental Management Program. This conference was jointly sponsored by the US Department of Energy. Edward Teller had originally agreed to give the keynote speech, however, shortly before the meeting he received an invitation from the USSR to visit Moscow and attend weapons-related meetings. He told me that it would be his first visit to Moscow, so he really wanted to go. In his place he asked Dixie Lee Ray, a close friend, to substitute for him. Dr. Ray gave an impassioned and controversial speech against global warming. In private conversations with me Dr. Ray expressed an strong interest in organizing a future meeting at Erice on the topic of radiation hormesis, a favorite topic of hers. Unfortunately, she passed away before we could organize the meeting.

The November 1995 meeting sponsored by the WFS International School of Innovative Technologies For Cleaning The Environment was the workshop, "Risk Management Strategies Applied to Environmental Cleanup in Central and Eastern Europe." This workshop was jointly sponsored by the US Department of Energy. Many scientists from eastern European attended and described the pollution problems in their countries.

The International Seminars on Nuclear War also provided memorable experiences for me. The August 1991 meeting of the International Seminars on Nuclear War was especially exciting, because of the attempt by a group of members of the USSR government to take control of the country from Soviet president Mikhail Gorbachev during August 19-21, 1991. I remember the USSR delegation huddled around the television in the computer room intensely watching CNN for the latest news of the coup, which was occurring during the Erice meeting.

Permanent Monitoring Panels on Planetary Emergencies were formed in 1998, and I was named the Chairman of the Permanent Monitoring Panel Pollution. Since then the Pollution PMP has organized numerous Seminar sessions and workshops involving many scientists from around the world, including workshops on Pharmaceuticals in Wastewater Used for Irrigation (1998), Black Sea Pollution (1999 & 2000), Caspian Sea Pollution (2001), Monitoring and Stewardship of Nuclear & Hazardous Waste Sites (2003), Groundwater Pollution Mapping in Sicily (2005) and Plastic Contaminants in Water (2006).

My most memorable experience was my participation in the March 1999 WFS delegation to the Vatican and our meeting with His Holiness Pope John Paul II. He was a very enthusiastic supporter of Professor Zichichi and the work of the WFS. The meeting was the 20th commemoration of an earlier 1979 meeting in the Vatican between H. H. John Paul II and a WFS delegation of scientists, which was the initiation of an unprecedented collaboration between Science and the Church. Each member of the delegation was personally introduced to the Pope, and during the introduction the Pope presented to me the Papal medallion. It was an honor to meet Pope John Paul, a memory I will always cherish.

In December 1999 I participated in the signing of a Memorandum of Agreement between the U.S. Department of Energy and the WFS. Theresa Fariello, Deputy Assistant Secretary for International Energy Policy, Trade and Investment signed the MOA for the Department of Energy, and Professor Zichichi signed for the WFS. The objective of the MOA was to initiate a framework for scientific cooperation and coordination on global environmental problems between the two signatories. This led to several joint workshops and Seminar sessions.

In June 2003 a small group of WFS scientists visited Italy's Gran Sasso National Laboratory at the invitation of Professor Zichichi to evaluate the hydrologic impacts on local water supplies of the Gran Sasso Laboratory extension. This group included including myself, Soroosh Sorooshian, Margaret Peterson, William Sprigg and Robert Clarke, all from the University of Arizona, USA, and Aaron Yair from the Hebrew University, Israel. A final report was generated and forwarded to Gran Sasso. This was another example of WFS scientists contributing their expertise to help Italy address its environmental problems.

Also in June 2003 I participated in a joint Rome meeting between eleven WFS scientists and Italian scientists to develop multi-disciplinary projects to improve the quality of the environment in Sicily. This effort was an excellent example of the power of the WFS to bring together environmental experts together with the local stakeholders to address and resolve local social problems. Projects were proposed for Sicily in the areas of water and environmental protection, floods, desertification, and seismology. I worked with Robert Clarke (University of Arizona), Aurelio Aureli (University of Catania), Lorne Everett (The

Shaw Group, US), Vittorio Ragaini (University of Milano), Luca Rossi (Italian Civil Protection Agency), Paolo Ricci (University of San Francisco) and Gina Calderone (EA Engineering, US).

As part of this important WFS effort I was directly involved in writing a proposal for the Sicilian Regional Resource Commission "Proposal for Pollution Vulnerability Mapping Database for Sicily". It was prepared in conjunction with Aurelio Aureli of the University of Catania, Massimo Civita (Italy), Giovanni Barrocu (Italy), and Salvatore Carrubba (Sicily) to develop a groundwater pollution vulnerability mapping database for Sicily. The goal of this project is to preserve the groundwater resources of Sicily for future generations and to prevent pollution that could limit the future growth of the Region's economy. The database will be made available to interested government officials and scientists via the Internet, and will become a model for similar developing countries with threatened and limited groundwater resources.

In December 2004 I participated in the first international meeting on the ILSEAT (International Laboratory for Science, Engineering and Advanced Physical and Biomedical Technologies) project at the Enrico Fermi Centre in Rome. The meeting also celebrated Fermi's main achievements and the CERN 50th anniversary. The ILSEAT project launched science, technology, planetary emergency, and cultural multi-disciplinary projects throughout Italy in which WFS scientists participated.

These personal anecdotes serve as one testimonial among many others to Professor Antonino Zichichi and his passionate belief that Science can serve mankind and his dedication in putting his belief into action. His historic accomplishments serve as a model for all scientists to emulate.

Zenonas Rokus Rudzikas

I have participated almost every year since 1991 in the famous Erice International Seminars. The first impression, the trip from Palermo airport to Erice by car, usually in the evening – ending late at night, proved impressive. For me, born and living in a country of lowlands (the highest hill in Lithuania is about 300 meters above sea-level) the hills and mountains of Sicily looked fantastic in the darkness of southern nights. Miraculous was the final part of the journey – a serpentine to Erice, the road sign "Erice – city of science", the view of the sea itself, as well as the myriad of lights far below the road.

Impressive were the opening ceremonies of the Erice Seminars in the Dirac lecture hall, with the introductory talks by Nobel laureates Kai Siegbahn, Tsung Dao Lee and, above all, the fantastic improvisations by Antonino Zichichi, known to friends and close colleagues – simply as Nino, with his charismatic figure crowning every event. Interdisciplinarity, crossdisciplinarity, synergy, global integrated vision, Planetary Emergencies with their unique classification – everything was new, exciting and astonishing.

Unforgettable was the opportunity to speak to the Father of the USA H-bomb Edward Teller, the American astronaut David Scott, and many other worldwide known personalities. And everywhere was the unseen presence of His Holiness Pope John Paul II …

Let me recall the beginning of the participation of Lithuanian scientists in the life of the Erice community. It started with the formal establishing of the Lithuanian Branch of the International Centre of Scientific Culture – World Laboratory (ICSC-WL) by the signing of the General Agreement between the President of the ICSC-WL, Professor Antonino Zichichi, and the President of the Lithuanian Academy of Sciences, Professor Juras Požėla, in Moscow on December 12, 1990. The relevant Decree (No 371) of the Government of the Republic of Lithuania was signed on December 11, 1990, in Vilnius.

Professor Juras Požėla invited me to assist him in the organization of the activity of the Lithuanian Branch of the ICSC-WL and appointed me its Director. My first visit to Erice was in August, 1991. At that time, in order to travel abroad we had to first go to Moscow. There, at the Academy of Sciences of the Soviet Union, we had to obtain a special service passport, as well as flight tickets, for traveling on to other countries. As a rule this procedure took several days and usually any free time was spent sightseeing. I went to the centre of Moscow, walked to the Red Square to look at the Kremlin surrounded by a high wall. The next day (it was a Sunday) I flew with Professor Juras Požėla and his wife to Erice, checking in the San Francesco Hotel late in the evening.

On the Monday morning I went to San Rocco where I met Professor Juras Požėla's wife. She told me of the rumour that something strange, but important, was happening in Moscow. However, no one had any reliable information on the political situation in the capital of the USSR, and telephone calls to the Soviet Embassy in Rome gave no particular results – the answer was that the Embassy had no official information. Moscow TV was broadcasting P. Tchaikovsky's "Swan Lake" and classical music …

Only a day or so later did we learn, from CNN and Moscow TV, that there had been a coup to throw Michail Gorbatchev out of power, and to stop Perestroika, which to some extent he had initiated. Fortunately the coup was suppressed and ended in shame. Particularly dangerous was the situation in my homeland Lithuania, where Parliament had already declared Independence from the Soviet Union on March 11, 1990. Professor Juras Požėla joked in Erice, saying: "We are living in communism (accommodation and meals during Erice Seminars were free for the participants) and watching a bad movie about socialism…"

I returned to Lithuania, again via Moscow. Once more I went to the centre of Moscow this time to see the traces of the coup and the remains of barricades. Something seemed strange to me when looking from the Red Square at the Kremlin. After a while I understood the reason for this – the former red flag of the Soviet Union above the Kremlin had been replaced by the three coloured flag of czarist Russia.

Arriving and staying in Erice was also a journey in time, going back to some extent to the Middle Ages. This feeling was due, not only to the age and location of Erice (some extremely narrow streets, and the uniqueness of the pavements and streets), but largely to local inhabitants and their way of life. Souvenir shops open until midnight, streets and squares crowded with smiling people, families with small children and babies. People either strolling, or sitting and chatting and simply just relaxing. And, of course, a view from the plateau facing the blue sea or picturesque mountains in any direction.

Unforgettable days and evenings, sometimes inside the clouds or above the clouds, being together during lectures and discussions in the Dirac, Feinman or Wigner lecture halls. Lunches and dinners in local restaurants or cafés, chatting, and sometime even singing in the Marsala room. Local folk song concerts and dance company presentations. I am very fond of singing. I have been singing in various choirs for many years, and it is why I listened to the singing with special attention, admiring the cultural richness of Sicilian songs and dances, the way of singing and the timbre of the voices.

Participation in the Erice Seminars, in the activities of the Permanent Monitoring Panels on global threats and Planetary Emergencies enriched every participant, and led to new ideas and endeavours. Under the direct Erice influence we initiated in Lithuania the first state financed national scientific programme, the "Ecological Sustainability of Lithuania in a Historical Perspective". It lasted for six years and united approximately 200 scientists from 30 institutions working together. After complex interdisciplinary studies of chemical and radioactive pollution of the soil, water and atmosphere in Lithuania and their influence on micro (genetic) and macro levels (flora and fauna, human beings, etc.), scientific conclusions and practical recommendations were formulated and presented to the scientific community and the Government of Lithuania.

In appreciation of the exceptional role of Professor Antonino Zichichi in the stimulation of a science without secrecy and borders his devotion to promote Science for Peace for the survival of our planet Earth, as well as his support of young researchers of our country, the Lithuanian Academy of Sciences elected him as its Foreign Member in 2005. We wish Antonino Zichichi – or simply – Nino, Member of the Lithuanian Academy of Sciences, the fulfillment of all his ideas for the survival of our planet and the welfare of mankind. Let us work together for his dreams to come true.

Hiltmar Schubert

My active participation in the Erice International Seminars over the last 15 years has enriched my scientific knowledge on the numerous issues the seminars dealt with. The international participation of members of different scientific disciplines makes this Symposium a unique source for creativity. The initiator, patron and promoter of these International Seminars on Planetary Emergencies is Professor Antonino Zichichi, President of the "World Federation of Scientists", who has served so well the international scientific community. Professor Zichichi first invited me to participate in these annual seminars on the occasion of a meeting of a subcommittee of the Scientific and Environmental Affairs Division of NATO, dealing with the approval of Advanced Research Workshops in the field of demilitarization after the end of the "Cold War" (He was chairman and I was the German delegate of this committee).

To deal with the different emergencies during the Erice Seminars Professor Zichichi created a framework subdivided into 15 classes: Water, Soil, Food, Energy, Pollution, Limits of Development, Climatic Changes etc. These classes provide focus special topical session plenary sessions during the seminar. I myself concentrated on two classes of interest: "The Huge Military Investments" and "Limits of Development". The item "Huge Military Investments" is a field of special interest dealing with demilitarization including dismantling of nuclear warheads, destruction of chemical weapons, and conversion of war industries. In starting to develop an overarching evaluation the huge investments in these munitions, we discovered the openly available, international database was both incomplete and replete with incomparable numbers. We, therefore, limited our attention to disarmament of munitions and conversion of war industries. Our conclusion was that demilitarization will neither be a profitable business nor will it yield a "peace dividend payout". As disarmament is expensive and as conversion to commercial products offers very limited possibilities, disarmament activities will still last for many years in many countries of the world. We discovered rather than the hope promoted in many promising ideas, conversion of war industry to civil activities will, in most cases, be an illusion. For example while the destruction of chemical weapons by incineration may come to conclude in some countries in the near future (USA), it will continue in other countries as an ongoing job (Russia, the Balkan countries, Georgia).

The other Emergency, which should be mentioned in this contribution is "Limits of Development". A "Permanent Monitoring Panel" (PMP) was set up for the Limits of Development and has made each year, up until now, many interesting contributions. The working group started with examining the limits of the development of megacities imposed by limited infrastructure aspects like water-supply, sewage systems and waste disposal management (1999-2003).

In the later years this PMP has been occupied with social aspects of migration (domestic and international), limits imposed by urban structures, and assimilation of immigrants. Following the trend to globalization a dynamic concept of "Limits of Development" was discussed according to new priorities like economic and climate changes. These considerations led to the principal subject "Climate Change" in different parts of the world and its consequences upon limits of development. Cooperation and joint meetings with the "Climate Change" and "Energy" PMPs were organized. Another priority was selected last year: "Sustainable Energy Supply" and its implications for limiting of development. Other interested PMPs were invited again to a joint meeting.

The collected published results of the PMP "Limits of Development" provide a useful international basis for present and further realization and use. The international cooperation of the members of PMPs is a benefit for the members themselves. The activities described above are representative of the activities of the PMPs that Prof. Zichichi proposed and promoted. The well organized general meeting presents interdisciplinary contributions concerning selected themes and the work of the relevant PMPs to the international community of scientists and for the participants themselves. We all have to thank Professor Zichichi for these possibilities.

At the end of this contribution an additional remark should be made. On one occasion of the Erice Seminars, I had breakfast together with Prof. Edward Teller, the "father" of the Hydrogen bomb. After a moment he questioned me on new research results on explosives and he amazed me with his knowledge on this topic.

Geraldo G. Serra

Arriving in Erice for the first time ten years ago, I had one of the most exciting intellectual experiences of my already long life, for many and diverse causes. First of all, the sight from the road of this huge marble quarry, in the middle of the sacred land of the old Magna Greece, made me think about how sustainable could be the use of the Earth's resources. Indeed, the Sicilian marble used to built walls, temples, monuments and simple houses from immemorial times, and the size of those quarries suggested that, with modern technology, they can attend human needs for many centuries more.

The ascent road to Erice comes just after letting us contemplate the immense landscape marked by human activities for so many centuries and that even more immense Mediterranean Sea, reminding us of the Ulysses odyssey and that land of Cyclopes. By the way: to reinforce my impressions, this evening welcome dinner was at the Ristorante "Il Ciclope"...

I don't want to be too literary, however, it is impossible not to mention my deep impression of those small tortuous streets, of the Piazza Humberto I and particularly of the three Ettore Majorana buildings: the San Domenico, the San Rocco and particularly the San Francesco, where I felt like a medieval monk looking for the transcendental Truth.

During August there three sorts of people in Erice: Ericini, Italians on vacation and the Erice International Seminars' participants. Being fluent in Italian, I enjoy very much any contact with local people, although mostly with those working in hotels, restaurants and shops, as well as the Ettore Majorana Center staff, who are so helpful. On the other hand, Seminars' participants form one of the most interesting and instructive gathering I ever saw. The emphasis in interdisciplinary approach, one of the Seminars' golden rules, brings together first order specialists in many research fields that have to dispense with professional lingo and interact with people of other fields, only to discover the existence of common structures to access reality in such different fields as physics, neurology or linguistics. This is enough to give an idea about the organization complexities Claude Manoli has to deal with every year to guarantee success.

Some of these researchers are members of my Permanent Monitoring Panel "Limits of Development" and with them we exchange mails all year along, preparing the annual meeting. With Manolo Borthagaray, Alberto Gonzales-Pozo, Chris Ellis, Bertil Galland, Hiltmar Schubert, Diop Mbareck and Leonardas Kairiukstis we worked hard for many years trying to find the path for a sustainable development and to identify obstacles opposing economic and social development in certain regions and countries.

Among the many superior minds I have met in Erice I would like to make particular mention to Kay Siegbahn and Antonino Zichichi. Kay Siegbahn, a Nobel Prize son of a Nobel Prize, invited me to his table at 1999's last dinner in Erice. It was for me a real remarkable occasion being the first Noble Prize laureate I had ever met. He was a real gentleman and gave me a very warm attention. Antonino Zichichi, who I have met at least once a year from 1999 onward, teaches us every year new aspects of the quest for natures secrets as well as the importance of a rigorous scientific posture confronting the problems of our time, particularly referring to its mathematical treatment. His defense of science against the accusation of being associated with weaponry creation, shows that weapons have preceded science for thousands of years and are ever associated with political decisions and technologic invention instead of scientific discovery. Science is only concerned with development of new knowledge about nature and not inventions of any sort, particularly that related with hatred.

Professor Zichichi's leadership is at the center of the Erice International Seminars which are difficult to imagine without his guidance.

Summarizing our findings in the search for a sustainable development would be too difficult, but maybe I can say that is almost never possible to use only renewable resources or, when this is not possible, to use non-renewable resources at such a pace that allows us enough time to come up with a substitute. Development is mandatory while a large part of humanity lives under sub-human conditions and without any perspective to break the vicious circle of misery. Therefore, one of the most important problems confronting science in our time is discovering the paths that lead to a sustainable development, particularly in facing climate changes and energy crisis.

These last years, in association with other PMPs we began to study the conundrum "climate change x energy crisis x development". This cooperation has permitted me to interact with other remarkable personalities such as Bill Fulkerson, Bill Sprigg and Richard Wilson. I think it was Wilson who called our attention to the Niels Bohr phrase: "Prediction is very difficult, especially about the future", which made me become more cautious than I used to be.

Although I have mostly monitored the Brazilian advances on ethanol technology, I have learned a lot about the perspectives of nuclear energy and climate changes, for instance. After all these years researching and discussing sustainable development I am convinced that new scientific discoveries and new technologic inventions can open new and unsuspected opportunities to economic and social growth. We don't need to limit human development but to boost it by scientific and technological research, finding new sustainable paths.

The Erice International Seminars, created and maintained by Professor Zichichi, have contributed, through thousands of scientific and technologic papers, to the better understanding of nature and to solve the above mentioned 3E conundrum: energy x environment x economy.

Soroosh Sorooshian

My very first exposure and familiarity with the Erice International Seminar series started with a brief hallway conversation with Dr. Robert Clark, my colleague at the University of Arizona back in the early 1990's. Bob, who had been a long-term participant in the World Lab and its various activities proposed the idea of my participating and sharing my experience on global water resources issues at the seminar. I was invited by Professor Antonino Zichichi a few months later and I attended my first seminar in August 1992. Among the memories that will never fade is the drive from Palermo Airport up to Erice followed by the check-in at the Ettore Majorana Center for Scientific Culture and the hospitality shown by the staff. The next was the introduction to the workings of Erice, including the unique procedure used by participants when dining in the local restaurants.

The opening ceremonies of the Erice seminar as a first time attendee was perhaps one of the most intellectual stimulating experiences in my career. As a member of the academic community, I had attended many meetings and conferences prior to the Erice seminar, but none had the degree of diversity of topics, ranging from climate change to nuclear proliferation, AIDS, threat of cosmic objects, etc. As a result, the first year experience was enough to engage me more seriously and continuously each year. My role in the seminar somewhat solidified as having been asked to lead the Planetary Emergency Panel on "Water". I had the pleasure of organizing a number of sessions on the interface of water resources and climate change and social impacts. Many distinguished scholars throughout the world accepted my invitation to participate over the years. Needless to say that the experience has always been the highlight of each year in the month of August. The consistency of the dates of the seminar (19-24 August of each year) has always allowed excellent planning ahead of time, as evident by the success of seminar series over the years.

The range of discussions at the seminar itself over the years, sometimes addressing controversial topics in a spirit of healthy discourse, has added extra value to the entire experience. The social events of lunches and dinners for the entire week has resulted in making new friends, learning new ideas, and sometimes acting as networking opportunity. A good example is my having played a role in bringing together Dr. Farid Hossani, Professor of Medicine at Fardossi University in Khorrasan, Iran and Dr. Franco Buonaguro, Instituto Nazionale Dei Tumori, Naples, Italy, which resulted in collaborative work related to AIDS research. I am not sure if they have ever met each other in person, but certainly the outcome of the collaboration resulted in a useful contribution in AIDS research. It is therefore these types of features of the Erice seminar that gives it uniqueness that is unmatched by any other.

Needless to say, that the success of any activity depends on the vision, tenacity, and belief of the individual(s) who initiate the activity. In the case of Erice's seminar series, without doubt, Dr. Zichichi and his remarkable foresight and commitment has yielded the success that the Erice seminar series has deservingly achieved. Of course, any good vision requires effective implementation. I would be remise if I did not acknowledge Dr. Claude Manoli and his staff, for their role in carrying out the vision set forth by Dr. Zichichi.

It is certainly a great idea that the impact of the seminar series be recorded for posterity and provide an opportunity to reflect on its history. For my small part, we brought together a number of world experts to sit together and discuss the emerging issues related to global water resources and stresses on the hydrologic cycle and the world's eco-systems. Over the years, the World Federation of Scientists established Permanent Monitoring Panels, including water. In the spirit of the world monitoring, we were able to assess the state of knowledge about regional water balances and evaluate the impact of climate variability on hydrologic consequences such as floods and droughts. Therefore, it was natural that the three working groups of water, climate, and extreme weather events, co-operate in bringing experts that appreciated the interdisciplinary nature of the issues facing our planet.

Through these interdisciplinary seminars over the years, it became clear, that as far as planetary emergencies are concerned, these three activities cannot be addressed individually but rather should be considered collectively. As the Erice International Seminar programme continues into the future, we will experience a lot more interdisciplinary approaches towards tackling major global resource issues such as adequate water supplies for the growing global community that are of the quality that sustain health and prosperity for all.

In closing my thoughts, it is appropriate to acknowledge an indirect, but highly important bi-product of my participation in the Erice seminars. Towards the later part of the 20th century, the issue of water was appreciated by the World Laboratory to the extent that it resulted in the formation of the World Laboratory Harshburger graduate fellowship program at the University of Arizona. Through the generous support of the World Laboratory, we were able to support nearly a dozen graduate students from the developing countries who came to the U.S. to study this important issue. All of the recipients of the World Lab fellowship completed their degrees and have occupied important positions in the international community, in academia, government agencies, and private industry. This could not have been possible without the Erice Seminar participation on my part and the willingness of Professor Antonino Zichichi to support this endeavor.

William A. Sprigg

Every year, for nearly three decades, Professor Antonino Zichichi steps to the podium in Erice and announces to a hundred or more participants in the International Seminar on Planetary Emergencies his faith in scientific "interdisciplinarity." Now, true interdisciplinarity is hard to achieve. First, it is a matter of time: time taken away from exploring ever deeper into one's narrow discipline. And a matter of value: recognizing how another's disciplinary assets can complement your own and recognizing potential for making a difference, either in one's research or in improving the human condition. The Erice Seminars have been the only consistent international forum to encourage and enable this on such scales of time and participation.

Through these seminars, and through the structure of Permanent Monitoring Panels, I have had the opportunity to promote interdisciplinary study of climate and the means to cope with climate variability and change. I have been able to participate in discussions ranging from terrorism and emerging infectious diseases to energy independence and environmental protection, with scientists, engineers and diplomats. I have witnessed the co-discoverers of the AIDS virus, Luc Montagnier and Robert Gallo, reach a meeting of the minds, putting to rest their controversy that had threatened progress in AIDS research. I have heard and debated proposals from Edward Teller that would alter the atmosphere in order to combat global warming. And, I have had the honor to co-chair technical sessions with Kai Siegbahn, T.D. Lee and Antonino Zichichi himself.

The Climate PMP prepared a position statement and white paper supporting free and open access to scientific data and information, which was endorsed by the WFS and presented to U.S. government agencies, the U.S. Academy of Sciences, the World Meteorological Organization and the World Intellectual Property Organization. Champions within the USG, US/NAS, and WMO used the arguments of the White Paper, developed in Erice, to pressure the international community to consider the scientific implications of governmental restrictions on access to data. The White Paper undoubtedly played a role in easing the restrictions that governments wished to impose.

When the Arab League requested assistance in constructing a plan for a dust storm prediction program, Robert Clark and I were called. We had the opportunity to work with the Arab League in improving their plan and promoting it at the WMO. Even though this effort failed to achieve its objectives, it laid the foundation of ideas for three subsequent successes: The World Laboratory-sponsored Malta School for Dust Forecasting; The WMO Sand and Dust Storm Warning Advisory and Assessment System; and the interdisciplinary NASA project, Public Health Applications in Remote Sensing.

Another tradition of the Erice Seminars and the genius of Zichichi is the invitation to dinner with all participants, usually around large tables. Here, many wide ranging discussions take place. On one such occasion I had the pleasure to sit with the science advisor to the President of Senegal, Mbareck Diop, who raised a serious problem concerning the safety of Senegal's fishermen. Along the Atlantic coast of Senegal, lacking electricity and many modern conveniences, fishermen launch small boats and head out to sea, where fishing is good but where unforeseen storms can capsize the small craft. Many lives were being lost each year. We devised a plan that evening to address the problem. We invited experts to the Erice Seminars the following year to demonstrate field radios that required no electricity. These were subsequently donated by the U.S. National Oceanic and Atmospheric Administration, distributed to fishing villages along the Senegal coast, and operators were trained to tune in each day for offshore observations of sea state and weather from hurricane research buoys and forecasts from the Senegal weather service. From ideas generated in Erice, fisher safety has been improved.

Climate variability and change, and their consequences, have generated endless debate during the Erice Seminars. It has been my pleasure to preside over these as chair of the Climate PMP. It is the most multi-disciplinary of subjects, so most participants in the Seminars feel free to contribute. These debates have spurred serious investigation of accusations of bias and restriction of data access. They have also led to multi-year collaboration among PMPs (Energy, Water, Defense Against Floods, Limits of Development, Pollution, Desertification, Infectious Diseases, and Climate), with notable results. A special session, "Framework for Managing Climate Change," was organized by the Energy, Limits of Development and Climate PMPs and held, by invitation of the Pontifical Academy, in The Vatican. This led to a series of papers that were published in the 2008 seminars proceedings.

Other research ideas, offshoots of the climate debates, concern climate implications in the outbreak and spread of infectious diseases. Franco Buonaguro of the Istituto Nazionale dei Tumori "Fondazione G. Pascale" and I discussed one such idea during the 2008 Seminars, and began examining the feasibility of a joint study of African dust storms, hitchhiking bacteria, and outbreaks of meningitis in Sub Saharan Africa. Such a study would need the engagement of atmospheric scientists, satellite data analysts, biologists, chemists, epidemiologists, medical researchers, and physicians. This is truly in the spirit of "interdisciplinarity," and of the Erice Seminars.

This is a personal memoir, incomplete, for there are many more examples of how I and others have been stimulated by the Erice forum. It is a place where scientists may meet with diplomats and practitioners of the arts that benefit from science. It is a place to air ideas and to enrich one's experience and exposure to different cultures and different priorities. Antonino Zichichi has brilliantly addressed the main reasons why people do not become "interdisciplinary:" Time, at least a week each year, participants are drawn away from the exigencies of lab and office; introduction to other, greatly divergent, science through the multi-disciplined Seminars; and exposure to planetary emergencies that require the attention of many disciplines in order to address them, through the continuity and activity of the Permanent Monitoring Panels.

Following each year's Seminars, I descend from the City of Science with more ideas than when I ascended the winding road to Erice. Each year I have but one regret: I haven't the time to chase them all. Perhaps I can pass some on to others.

Bruce Stram

The World Federation of Scientists meetings are of course remarkable for the intellectual capacity assembled and resources that are brought together. The interactions that occurred there related to nuclear arms, disarmament, and arms proliferation are legendary.

I was not privileged to be in attendance when most of those events occurred. However, I can attest to the continuing importance of Erice in other areas. At the first meeting I attended in the mid nineties, we were treated to presentations by the two major aids researchers, Robert Gallo and Luc Montagnier. They were then at odds. They each claimed to have discovered the AIDS virus. They were viewed as rivals.

It was therefore a coup to have them at the same place in a non adversarial environment. But Professor Zichichi took it one step farther. He prevailed on the two men to join together on stage, and with a handshake, agree to focus greater research efforts on remediating the Aids plague in poor nations. The rivalry was complex and commercial issues intruded, but they did ultimately "bury the hatchet".

A number of years later, Erice provided a forum for a group of young scientists to expound on their efforts to develop "pharma" plants, which offered the promise of using plants to produce vaccines effective against human diseases at fractions of the cost of existing methods. They could not help but be encouraged by their opportunity to present and attend at Erice. On a personal note, my wife and I sat with a group of these researchers at a dinner, a thoroughly enjoyable experience.

We were also delighted to have met Doug Morrison, whose quick mind and range of topics was impressive. He had well supported theories on the Cretaceous extinction and kept us abreast of cold fusion developments (or the lack thereof.) His premature loss was painful.

Jan Szyszko

"Neither science nor civilization could exist without memory" (Antonino Zichichi 2007). I met the author of these words on April 27, 2007 in the Vatican at the "International Seminar on Climate Changes" led by cardinal Renato Raffaele Martino, the chairman of the Papal Council "Institia et Pax", where I presented a lecture entitled "Climate change and state responsibility". Our first contact made a great impression on me. I met a man with a precise mind, at the same time permeated with deep humanism. He came to my lecture and could not listen to it because I gave it one hour earlier than planned.

The lecture was transferred to an earlier time because I had to be in Luxembourg at 3:00 to represent the Polish Government as the Minister of the Environment at the meeting of the European Court of Justice. It was there that a hearing against the Polish Government concerning the motion submitted by the European Commission was to take place according to a speeded-up procedure even though it was supposed to be held at a much later date than the Vatican Seminar. The motion referred to an immediate suspension of compensation activities relating to the planned deforestation of about 20 hectares of the forest within the defined limits of a Nature 2000 area, occupying more than 150 thousand hectares, necessary to build the Augustów bypass (north-eastern part of Poland).

Compensation activities were to involve the afforestation of 100 ha of the adjoining area occupying 7156.9 ha, to be determined (according to the European Commission), also within the Nature 2000 network. The reason why the motion was filed against Poland was the fact that such priority Nature 2000 species like the pasque flower and bractless toadflax occur in the suggested Nature 2000 area. According to the claimants, the afforestation of 100 ha could have a negative impact on the occurrence of these two species. The hearing in Luxembourg revealed the dullness and ruthlessness of the EU bureaucracy with no place for ethics or science, i.e. for truth. At the hearing against Poland in the European Court of Justice, the Polish minister said:

- Poland cannot stop the compensation (afforestation) process because it has not commenced it.

- Neither the pasque flower nor the bractless toadflax occur in the plots to be afforested, which include arable lands lying fallow for a few years or illegal refuse dumps. Maps prepared with the exactitude up to 1 m in the presence of Polish professors of the global scientific authority were submitted as an appendix.

- It is a paradox that the possibility of occurrence of the pasque flower and bractless toadflax has to be preceded with the planting of a forest and, in the case of the pasque flower, the forest has to be used in an appropriate manner, strongly established in the culture of the local communities in those areas.

I mention it as an anecdote because it was the reason why professor Zichichi could not listen to my lecture and began to look through my manuscript in my presence. In the introduction, I wrote: "According to many scientists the increase in the content of carbon dioxide in the atmosphere causes unfavorable climatic changes, thus contributing the global warming. This increase results from the combustion of carbon accumulated in hard coal, brown coal, oil, gas and timber.

It is also connected with land use. In Central Europe (Poland) in the natural forest exist about 300 ton of carbon per hectare while in degraded forest or previously cultivated deforested areas about 20 ton per hectare exist. I also said that "For the sake objectively it has to be noted that many scientists express a different opinion. They believe that climate change and climatic anomalies are natural phenomena. For example, in central Europe warming following the last glaciations began several thousand years ago.

The current period saw considerable warming as well as cooling that had no connection whatsoever to human. I also wrote there that "Regardless of whether global Warming poses a real threat, the important fact is that in 1992 the Framework Convention on Climate Change of the United Nation came into effect, expressing the political will of the majority of governments and parliaments in the world. The basic goal of the convention is to stabilize the concentration of CO2 in the atmosphere at a level that – according to the first group of scientists mentioned above – would prevent the dangerous anthropogenic interference with the global climate system. Hence, it is material that the Climate Change Convention can be used to create sustainable development of the world, where sustainable development is nothing else but fast economic development combined with the rational use of environmental resources and respect of human rights. It will be implemented only if we measure it with the use of indicators such as: increase in the gross domestic product, decline of unemployment, longer life and predominance of young classes in human population, water quality improvements, reduced emission of carbon dioxide and no disappearance (return) of native species.

I felt that professor Zichichi understood me and, for that reason, I was very happy to accept his invitation to Erice in 2007. I spent one day there and had a possibility to give a lecture indicating that the main problem of the developing world is in the mass disappearance of native species of plants growing wild and animals as well as an increased likelihood of the mass appearance of pathogenic organisms dangerous not only for the production but also directly for man, such as the avian flu, foot and mouth disease or mad cow disease. I indicated there that man and his mindless economy in space is responsible for them, causing the impoverishment and reduction of the differentiation of organic carbon content in the landscape. I also indicated that, if man with his economic activity can destroy environmental resources, he can also use the economic activity to make up for the mistakes and the Climate Change Convention could be an excellent instrument for that purpose.

During that short stay, I got familiar with the atmosphere of the scientific circles of Erice. The circles up to the highest scientific standards, discussing ethics in science and its role for the creation of global peace with utmost concern, treating the role of the ecology and hence the rational use of environmental resources by man as equal to the security of the energy, foodstuffs or nuclear forces. The words spoken by one of the greatest moral authorities of the world: John Paul II, winner of the "Prize Ettore Majorana – Erice – Science for Peace" award, who said in his proclamation on the XXIII World Peace Day on December 8, 1989 in the Vatican: "In our times, the awareness is growing that the threat for world peace is not only in the armaments race, local armed conflicts and various forms of social injustice still present in the life of peoples and nations but also in the lack of proper respect for nature, chaotic exploitation of natural resources and gradual deterioration of the quality of life.

Such a situation creates the feeling of temporariness and uncertain future whose consequences include many forms of collective egoism, greed of wealth and the application of violence.

Faced by the prevalent degradation of the natural environment, the humanity realized that it was no longer possible to use resources on Earth in the same way as it used to be done in the past. This phenomenon arouses the concern of the public opinion and politicians while experts in various scientific fields study its causes. This way, ecological awareness is being created, which should not be suppressed – on the contrary – we should promote its deepening and maturing so that it could find its expression in specific programs and initiatives".

I was able to feel the creation of such ecological awareness in Erice, walking the corridors and rooms in which the Great Author of the above-mentioned words stayed in May 1993, and these words were being implemented there with Professor Antonino Zichichi as a catalyst of the process. It depends on us whether it will be reflected in specific programs and initiatives. I can only say that "Neither science nor civilization could exist without the memory of the Science for Peace Erice Prize". I will never forget that I was the second Pole and the first ecologist to receive the "Prize Ettore Majorana – Erice – Science for Peace". That *noblesse oblige* just like truth in science. Because of that, I believe that programs such as "Sustainable human development and global change: landscape management in rural areas and greenhouse gases mitigation" will be based on the ethics and knowledge and not particular business interests or short-term political goals causing the lack of proper respect for nature, chaotic exploitation of natural resources and further deterioration of the quality of life.

Gerard 't Hooft

The top of a moderate size mountain, on the North-West of the island of Sicily, is covered by the small medieval town of Erice. With its roots in ancient history, way back to the times of the Greek storyteller Homer and the historian Thucidides, its streets unchanged since medieval times, and surrounded by ruins of the different ages, Erice attracts visitors from all over the world, who wish to breathe the atmosphere of the past.

But it is not only the past that makes Erice special; Erice is also very much involved with the present. Its visitors may be surprised to notice that several of the ancient cloisters, such as one called San Rocco, have been renamed after twentieth century scientists, and apparently serve as modern science institutes. Several dozens of Summer schools take place here every year, and scientists, ranging from young students to senior erudite professors, with nationalities from all over the globe, buzz in and out. Nearly always there are world renowned experts among them, including various Nobel laureates. Thus, besides being a medieval town, Erice is a center of modern science, where cutting edge scientific findings and discoveries are being scrutinized at the highest possible level. To a large extent, this is the doing of one man who is unquestionably Erice's most famous inhabitant, Professor Antonino Zichichi. Born very close to this little town, Nino knows the place inside out, and he injected into it his zeal for scientific investigations and discovery, with his passion for scientific excellence and grandeur, as much as his curiosity for the inner workings of Nature. Being a superb experimental investigator of the sub-atomic world himself, Nino also knows the world of science like nobody else does.

Atoms can be split and fused, this we know from the midst of the 20th century. But the material that the atomic nucleus itself is made of, its deeper relationships with the electrons that circulate the nucleus at what looks like relatively great distances, the rules by which these forms of matter move and interact, the processes of creation and annihilation that take place at tremendous rates inside the nuclei, turned out to be tremendously complex. It is here that Nature poses magnificent challenges to humanity. Nino and an impressive number of his scientific friends have accepted these challenges, and they are devoting their lives to discover answers. To discover how Nature works, one first of all needs to do experiments, try to create the most exceptional and unique circumstances for the tiny particles to be exposed to, and design the most sensitive detection devices to register their responses. In practice, this means that one has to construct the most powerful possible particle accelerators, and gigantic electronic machines filled to the rim with circuitry dedicated to do the most accurate possible measurements. Next, and besides, the utmost is being asked of our intelligence, our mathematical skill, to find the proper language that can be used to describe these particles, to visualize precisely what happens in a world that is extremely unlike the world of our daily experiences.

The official name of the Erice Institute is the "Ettore Majorana Foundation and Centre for Scientific Culture". Indeed, this centre is big enough to house many other sciences besides that of the sub-nuclear world. Sciences are being studied such as Medicine, Biology, Astrophysics, Solid State Physics, and the science of "Complexity". Each of these has the intense interest of Professor Antonino Zichichi.

At the interface of science and culture, many aspects of their mutual interactions are manifest, and indeed these are a continuous concern of Nino Zichichi. Only rich cultural environments can nourish and stimulate scientific curiosity; the people in a community must feel the need to pay for the expensive investigations, and it is obvious that they will want to see the benefits of their investments. Their rewards are the beautiful and intriguing stories scientists can tell them about the material that our world is made of, how our world came into being, what our place is in the Universe, and which natural and manmade changes lie ahead of us.

Our planet is continuously changing. Mankind is exploiting its resources at a fast increasing pace. Disasters of various kinds are continuously threatening us. Will there be war, famine, severe shorts of energy, food, water or medical care? Will climate change take place, how, and why? If human activities are among the causes, what kinds of actions, if anything, can scientists recommend to be taken? How should we respond to the continued threats of terrorism? Should our ever strengthening power of reason not enable us to confront the numerous difficulties faced by humanity in its perpetual struggle for peaceful coexistence? These are the kind of questions that are also being posed, and answered, between the walls of the ancient buildings of Erice.

Participants at the various scientific schools are shown the “Erice Statement”. This statement was written in Erice, August 1982, by Paul A. M. Dirac, Piotr Kapitza, and Antonino Zichichi. The Statement is a general appeal, both to individuals and to governments, to encourage the peaceful pursuit of science and to ban categorically all scientific activities aimed at the construction of weapons of mass destruction, in particular nuclear ones. Also, secrecy of scientific investigations for defense purposes is demanded to be minimized. Most scientists wholeheartedly agree. During subsequent years, the Statement has been signed by thousands of scientists, the world over.

It is not always easy to sustain the “grand View” when sometimes more mundane matters appear to ask more immediate concern. Should we defend complete freedom of scientific exchange everywhere, or should we support a ban against one or another malevolent government? Should we agitate against any employment of military power everywhere, or should we support and encourage small minorities to defend themselves the way they see fit? Scientists are people like all the rest of us, and the most ardent questions generate divides also among them, but they are fully equipped to address questions such as these and to analyze the complexity of these issues in a manner that is as objective as is possible, doing this just in the way they analyze and interpret complicated tracks of elementary particles in an electronic detector.

It is time to think about the future of Erice. Its tradition, as developed by the visionary founder of the Ettore Majorana Institute, must be allowed to transcend the fugitive times. The numerous stone buildings in Erice will certainly last for many more centuries, but for this profound scientific bulwark and unique cultural centre, we must do all we can to ensure many more years of fruitful activity.

Evgeny Velikhov

The book of testimonials about the Erice International Seminars and related endeavours that have been occasioned by Professor Antonio Zichichi's 80th birthday provides an excellent opportunity to recall many significant events, and to say a couple of words about the importance for the international scientific community of the World Federation of Scientists and the World Laboratory, both of which wer founded by this outstanding academic.

In the beginning was the Erice Statement, published in August 1982 by Nobel Prize winners and famous researchers Paul Dirac, Pyotr Kapitsa, and Antonino Zichichi. A good team of like-minded persons! This brief document written in a lapidary, almost mathematical language set the objectives and goals for World Science, emphasized the responsibility of scientists towards mankind, and outlined the prospects for the development of modern civilization, wherein state-of-the-art technologies may be both instruments for peace and engines of war.

This Statement, the World Federation of Scientists and the Ettore Majorana Center for Scientific Culture, which was founded by Zichichi as early as 1962 in the small Sicilian town of Erice, were catalysts for the creation of the World Laboratory. In this institution researchers from all over the world, regardless of their affiliation, work together and implement projects for the benefit of mankind. In my opinion, it was one of the most productive ideas of the 20th century with respect to consolidating the endeavours of the academic and broader intellectual international community.

It is no wonder then that the ideas of the Statement attracted the interest of and obtained support from many political leaders in the USSR, the USA, China, Italy, and Canada and renowned researchers. The concepts underlying the Statement and the practical activities at hundreds of international scientific meetings worldwide made a tremendous contribution in dismantling the Cold War philosophy and in fostering cooperation between nations; they also focused attention on the global challenges related to the development of human civilization that range from the impacts of nuclear catastrophe to the asteroid threat to the planet Earth. At the meetings held under the auspices of the World Federation of Scientists and the World Laboratory, specific scientific problems are discussed, bold hypotheses are proposed, and the programmes for the development of mankind in the remote future are shaped. For all these endeavours to be materialized, Antonio Zichichi, his vision and vigour were needed.

The World Laboratory is of real importance: when the USSR Division of the World Laboratory started operations, for Soviet scientists, it was the first really free research institution in the country. It helped a lot in fostering academic and human contacts and in promoting international cooperation. Undoubtedly it has made world science more open.

Many of my personal recollections are related to the World Laboratory and to the annual August sessions at Erice. I was lucky to have many amazing meetings with different people. I still see in my mind Edward Teller, the father of the US hydrogen bomb, vividly arguing with young Soviet physicists. The neighbouring building hosts a state-of-the-art seismic laboratory while in a cellar, whose stone vaults were constructed hundreds of years, ago one can find a barrel of good Sicilian wine and an old piano whose music entices you to sing and think.

And to conclude: Thanks a lot, Antonino, you have done a great job!

Henning Wegener

I was first invited to the International Seminars on Planetary Emergencies –devoted to Nuclear War – in Erice in 1984, a quarter century ago, and have come back again and again. Erice has become an inextricable part of my life and intellectual growth process. Whenever I think of these intense annual encounters in August, I do so with emotion and immense gratitude.

To start with my accumulated and growing debt of gratitude, it is obviously owed to Professor Antonino Zichichi as the creator and leader of the Erice universe. As a non-scientist, I often felt that the recurring invitations addressed to me were undeserved, as I had understandable doubts about my ability to contribute meaningfully to the natural science debates. But there was always a great amount of generosity and hospitality (the latter including Mrs. Zichichi's), and of warmth that immediately welded the seminar participants into a collective entity of friendship and common purpose. Nino is not only a great scientist, but a leader and organizer with unique charisma. He has been the unifying force of every Erice assembly, recruiting the best minds from global science, integrating and crystallizing their debates under the demands of unrelenting scientific rigor, and with a unfailing intuition for topics of highest relevance and consequence. Nino also stands for the attraction and significance of the Erice venue; his family home, - Sicily, Erice, uniquely embedded in European and Mediterranean history, a crossing point of civilizations and creeds, a place of magic beauty. One was always aware of the inspiration that emanated from this singular environment.

During the first years of my attendance, I drew direct professional benefit from my participation. First recruited as German Ambassador for disarmament affairs, and representative at the Geneva Conference on Disarmament and other disarmament bodies, then from the perspective of my work at NATO, the nuclear debates of the eighties, disarmament issues, non-proliferation, new weapons developments, nuclear winter, star wars and all, and later the slow diminution of East-West dangers and tensions, were directly relevant to my official endeavors, the more so since the dialogue with major Soviet - or Chinese, Arab etc. - participants was so uniquely fluid and constructive, open and relaxed. Polemics had no role in the catalytic environment of Erice and under the imperatives of scientific debate, in sharp contrast with the battles in UN assembly halls and tough intergovernmental negotiations . . .

Later, with the broad perspective on Planetary Emergencies, the learning process also grew broader, encompassing insights into the fragility of our planet and our societies. As in the scary age of potential all-out nuclear war, Nino had us concentrate on the evolving issues of supreme importance for the future of humankind, and on the contribution science could and had to make. For me, many of the sessions were eye-openers, personally enriching. Each August session, eagerly awaited and intensively lived, became a refresher course for the really important developments and perils in the world around us, top actuality delivered by the brightest minds of our time.

Rich rewards could always been drawn from the pluridisciplinary approach so characteristic of Erice. Unnecessary to emphasize, as a diplomatist, lawyer and social scientist - although with some academic ambitions - my fascination by the scientific discourse to which I was exposed was in most cases not underpinned by adequate understanding of the frightening series of equations and diagrams. Often I thought that the impressions of Charles Percy Snow – a feeling both of loss and admiration – when he in 1959 developed his theory of "The Two

Cultures" – the "Humanities" and the natural sciences – which would never understand one another and engage fully in dialogue, must have been very similar to my initial reactions. But C.P. Snow himself overcame this stage of thinking, and supported later by John Brockman, saw a Third Culture emerge which today culminates in an intense cognitive and conceptual dynamics between humanities and science, helped by the growing articulation ability and communication tools of scientists and the evolving digital age. Happily today, the transition between the disciplines is fluid, the sharp rift between the Two Cultures has been bridged. I have been privileged to observe this ongoing process in an exemplary environment, and with enviable opportunities for personal networking with fascinating interlocutors.

When in 2001 we embarked on our work in the Permanent Monitoring Panel on Information Security, another Planetary Threat, the value of an interdisciplinary work mode again became utterly manifest. Creating order in cyberspace, and elaborating recommendations to that effect needs a conjunction of scientific, technological, economic, legal and administrative expertise. On another aspect of this, although following the complex scientific debates on many topics with great interest and enthusiasm, I often tried to make up for my deficiencies in scientific knowledge and judgment, by trying to embed the debates in the diplomatic and legal frameworks which to me appeared necessary, and thus to inject an element of pedestrian political reality in the visionary discourse that was taking place, - presumably with limited effects.

At the same time, and more importantly, I learnt to admire the eminent scientists whom I heard and with whom I could converse. I admire the Galilean rigor of their scientific methodology and their work ethics, their infinite quest for truth and evidence, - but also their vision, their belief in what science can achieve, their courage. At one session – I think the speaker was Professor Lowell Wood, the subject one of the Teller-Wood Star War devices – one participant remarked that the findings displayed were still three orders of magnitude off the desired operational level. The Professor's reply: "But my dear colleague, three orders of magnitude is trivial!" This unshakeable belief in the validity of a non-falsifiable scientific premise left me with a profound impression. The Professor's conviction was only one expression of the shared belief in Erice that science is an indispensable tool for the resolution of global problems and conflicts. This belief in the role of science is vital, and indeed grandiose.

From the many remarkable attendees at Erice during the course of all these years, let me offer a small recollection of only three, all three of them recipients of the Erice Peace Prize.

Ambassador Qian Jiadong, my then colleague in Geneva, is fortunately still with us. He was responsible, jointly with Professor Tsung Dao Lee, for the early opening of Erice to China, and soon brought to the Seminars representatives of the highest echelons of Chinese science, an accomplishment that very much enriched our perspectives at the Seminars. But apart from that, I would like to pay tribute to him as an excellent, polished diplomat, a person of infinite courtesy and elevated personal culture. In his refinement and his human qualities, he has always appeared to me as the personification of the ancient and noble culture of his country.

The other two eminent persons here remembered, have passed away, like so many renowned Erice participants who have left us during the last quarter century. Most of these are still universally present in their writings and scientific achievements, but Erice is very much about personal encounters, character, moral imprint, human attributes. Personal recollections can help to preserve this precious legacy, so much a part of the Erice community and comradeship, and may give us the illusion that these great men and women are still with us.

Edward Teller is most vividly remembered by most of us. His overwhelming intellectual presence and universal grasp of problems, coupled with an exceptional kindness and personal nearness are legendary. My contacts with him are unforgettable to me personally, but may also display a trait of more general validity. Hungarian and American, Teller still seemed to be very attracted by a German interlocutor. I was always surprised on how many occasions – mostly of course in Erice, but once also at the Munich Wehrkunde conference, where he spoke about Brilliant Pebbles – Teller sought me out to engage me in conversation – in German!, and with a slight, but enchanting Hungarian accent. He was intensely interested in anything German: politics of course, culture, science, the ways of life after Unification, and including my personal views. I understood that this interest hailed from his years in Germany from 1926 to 1934, at the Universities of Karlsruhe, München, Leipzig (doctoral work with Werner Heisenberg), and Göttingen (with James Franck). Despite his forced exit from Germany, it appeared to me that his intellectually formative years there had left him with an important and almost tender souvenir which, perhaps, was even growing stronger with advanced age.

I would also like to recall my conversations with Sir John C. Eccles, the great Australian neuropsychologist and Nobel. With him and Lady Eccles I would spend long moments discussing a vast array of problems, commensurate with Sir John's immense knowledge and unbounded interests. In his untiring quest for extending scientific discourse to values and the spiritual dimensions of man, he was perhaps the ideal personification of the pluridisciplinary approach at Erice. But lastly our exchanges would necessarily recur to his life-long preoccupation with the human brain, the mind-body relationship, the human mystery that the mind included a part existing distinctly and unexplainable, distinct from its physical basis. Sir John brought home to me the miracle of human evolution against a myriad of odds, a "hazardous and tenuous evolutionary route" pursued over millions of years, as he wrote in his marvellous book "Evolution of the Brain: Creation of the Self". And he also wrote: "I believe that there is a Divine Providence operating over the materialist happenings of biological evolution", placing the spiritual aspects of human nature forever beyond the scientific explanations of evolution. For him, the unique essence of the human self could never be conceived without faith in a divine creator. I felt then, and feel, immensely comforted by these conversations, comprehending that science and faith -and here, evolution and creation - , are by no means contradictory, but indeed reconciled. As of this writing, we are commemorating the two centuries of Charles Darwin's birth, and it is good to celebrate his scientific genius along with the spiritual complement his discoveries require. In this optique, neither we – nor God – need the new forms of popular creationism that are traded in so many quarters today. Sir John Eccles' books describe in a moving way where we come from, and what we are. I owe him much, as I owe immensely to Erice.

Jody Westby

It is the hope of many that when they depart the earth, they will have made a difference. What makes a difference? Recognition in a chosen professional field? Influencing the direction of a child? A scientific discovery? Writings that spark a school of thought? Restoration of an ancient village? Advancing one's country? Bridging two disciplines, such as religion and science? Helping solve the world's most critical problems? Yes. Even just one of the foregoing makes a difference. Antonino Zichichi has done them all – and he continues to make a difference. He is a mega-leader.

Mega-leaders are visionaries and have an innate sense, an instinct about circumstances that are worthy of global attention. They also understand that today's problems are too interconnected and complex to be addressed by one discipline. Although WFS was formed by a group of scientists, Nino recognizes that today's problems must be approached from a multidisciplinary perspective. I saw the term "multidisciplinary" come to life most fully during a plenary session on terrorism. Policy, rule of law, and science came tumbling out in the plenary hall, mixed with personal opinion. The hall was vibrating; it was the most animated session I have witnessed in Erice. Many of the participants were unhappy about the U.S. invasion of Iraq in 2003 and were openly critical of American foreign policy. Just when it seemed that there was virtual unanimity on this point, Andrei Pointkovsky took the floor and forcefully defended the U.S. government, punctuating his statements by pounding the table! The Russians were defending the Americans!

The participants coalesced when Hussein al-Sharistani, a nuclear chemist who was imprisoned and tortured in Abu Ghraib prison for refusing to develop weapons of mass destruction for Saddam Hussein, sought the sought of the Erice scientists for the restoration of the Iraqi National Academy of Sciences. The scientists, policymakers, academicians, and lawyers in the room quickly united, huddled over laptops, shuttled drafts from room to room, and adopted a Statement of Support of Iraqi Scientists that was sent to major scientific academies. As a result al-Sharistani got support from the Royal Society and a specific meeting was held. Richard Wilson arranged a formal support from the American Academy of Arts and Sciences, but despite the efforts of Wilson and Dick Garwin, the US National Academy was lukewarm. Unfortunately, the situation in Baghdad soured, and there was no funding for an Iraqi Academy.

Nothing that happens on the floor, however, compares to Nino's opening plenary address. For me, it is the highlight of every year. What will his vision be this year? What will he say that will ignite conversation, arouse discussion, and stir thinking? I will always remember the first time I met Nino, the twinkle in his eye, the energy that surrounded him; it was something greater than charisma.... something more like a force. Then came his talk at the plenary session of the World Federation of Scientists, and I realized that I had met someone special....someone who could shape the thoughts of men and women, channel resources toward problems – Planetary Emergencies – and impress urgency upon governments. Nino has the ability to put his finger on such topics and moments in time, to understand their importance, and to motivate others to seek solutions to them. Then he quietly takes care of the hard part...raising funds to keep everyone engaged, bringing us back together, ensuring camaraderie and intellectual stimulation, and publishing our work.

There are also surprises, such as Edward Teller's last video conference in the plenary hall. We were arguing about Iraq and terrorism, and this scientific giant was worrying about meteors exploding and destroying life on earth. He was disappointed – and frankly, I think dumbstruck – when he asked if anyone had any questions and there was silence. And, then there are the other lessons learned in Erice…."discussion is more important than lunch"…"do not touch the microphone while speaking!"…."make proposals"……"WooHoo!"

Nino's role in establishing the Ettore Majorana International Centre for Scientific Culture is an additional proof of his leadership. In the process of turning an ancient town into one of the world's most renown scientific educational centers, he saved a Sicilian jewel that yet preserved its…. magic. The memory of the first time I sat in San Francesco, listening to the Sicilian singers and dancers as the fog wafted over the top of the courtyard walls has never faded. I can picture the white doves nestling in the eves and doorways of the duomos, the Roman roads of stones weathered to a smooth, waxed surface, the blue Madonna, and the lights of Trapani twinkling beneath a full moon hanging in the Sicilian sky. Wasn't it just yesterday? And that panoramic view from the coffee room....

As Erice's streets have become more crowded with tourists, I do not consider it a nuisance, but rather an affirmation of Nino's success in rescuing this enchanting mountaintop town with its Norman castle. How lucky we are to share it and enjoy Nino's gracious hospitality!

The intensity with which Nino approaches science is balanced by his warmth toward family and friends. In fact, it is quite instructive. More scientists should share the cover of their books with their granddaughters! The camaraderie of every August is due, in part, to the careful attention that Nino gives to the spouse program, their inclusion in the plenary hall, and the spirit of friendship that is encouraged at evening dinners… and in the Marsala Room! In all, the intellectual stimulation, beauty, congeniality, and passion that converge in Erice are unique, precious, and humbling.

Nino's leadership has made an enormous contribution to science, culture, knowledge, and art….and to my personal and professional development.

Richard Wilson

Each of the Erice scientific schools and seminars have tended to concentrate on a particular field of science. Usually expert practitioners in these fields are invited to discuss the status of the fields and explain the discoveries to the less expert. In these seminars there is of course the usual demand for rigorous scientific proof of any statement, with an exception that conjectures may be discussed in the context of searching for methods of proof or disproof.

It is of course desirable to maintain the same rigorous standards in what I call, for want of a more precise term, scientific policy seminars. Policy should be based on facts in so far as possible. All too often, important and far reaching policies are based upon fantasies or worse: on statements that can be or have been disproved. Although the desire for rigorous proof must remain, it must be recognized that, to be useful, policy statements can be and must be made with a lesser standard.

Discussion of the effectiveness of the seminars must be treated with even more caution. Although I firmly believe in some of the comments I make below, they are far from proven. There is no possibility of repeating world history without the Erice seminars. In astronomy similar alignments of stellar objects repeat themselves and this allows scientific rigor.

The first of these seminars was a series in the 1970s on Prevention of Nuclear War. I remember in particular meeting Nobel Laureates Pietr Kapitsa, Paul Dirac, Eugene Wigner and scientists in semi political roles, Edward Teller, Yevgeny Veliko, Richard Darwin and Frederic Sitz. I remember swimming at the beach below Erice with the inevitable anonymous KGB man. Intense arguments took place between Garwin and Teller about the "Star Wars" program with Velikhov a fascinated observer. I remember sitting at a small table for dinner at "La Pineta" with Velikhov, Seitz and the KGB "minder". After suitable lubrication Velikhov suddenly made the dramatic comment: "I hope that my country will soon abandon its intention of conquering the world." Seitz and I were nonplused. Today I might respond by saying that: I hope that my country will soon abandon its intention of dominating the world." Since then Velikhov confided to me with pride "I have had an audience with the Pope 6 times" - no doubt arranged by Nino Zichichi. Although he has never stated this himself, I believe this was important in his organization of the 1987 Conference on a Nuclear Free world in Moscow which helped to end the cold war. Many of us have the (unproven) belief that scientist to scientist contact was crucial to the prevention of nuclear war and ending the cold war. Nino gave us a focal point. The Pugwash conferences won the Nobel Peace Prize, but Nino's meetings may well have been more important. Alas, we have not yet achieved a nuclear free world.

The seminars on Planetary Emergencies which are patterned on the (presumed) success of the seminars on nuclear war may seem of lesser importance but one cannot tell. The dropping of the Atomic Bomb on August 6th 1945 marks the time when mankind knew how to destroy itself. Mankind is still struggling to find a way in which such destruction will not occur. I would like to see more emphasis on this in Erice, and in particular on the interactions, if any, between this struggle and other aims of society. This role, which was particularly discussed in Harvard Physics department when I first arrived 54 years ago, largely due of course to the two giants, Bainbridge and Ramsey who had been personally involved in the August 6th 1945 action, now tends to be discussed in schools of Public Policy or Government with less active scientific input.

The Erice seminars on planetary emergencies, address another aspect of civilization. Can we make whatever adjustments to nature that are necessary to prevent mankind being accidentally maimed or destroyed? The Black Death in the 14th century killed, perhaps, 1/3 of the human race. No one knows why it stopped. We clearly want to do better the next time and our understanding of the plague will enable us to do so. But there will be other threats of which SARS and Bird Flu are recent examples. Will energy use, which has brought so much prosperity to the world cause such dramatic changes in the world's climate that this prosperity is nullified, or even worse, the basis of our civilization damaged? We need not only to use our imaginations to calmly envisage possible catastrophes but our brains to analyze their possibilities. It is hard to discuss these in ordinary public fora. Mankind seems to be programmed to demand definite answers. Probabilities are not well understood and conditional probabilities less so. Scientists have to learn how to answer the questions: Is this chemical safe? Is it possible that people will die from global warming in the future? Without either being too definite, or too confusing. These issues may seem to defy scientific calculation. Rigor there cannot be. But usefulness there can be. I remember that the Canadians, looking at birth rate data, made arrangements for an increased number of births 9 months after the Montreal world's Fair of 1967. Open discussion of these issues in the Erice seminars enable us to go out in the less scientifically oriented world.

One of the most contentious issues is now that of global climate change which has been extensively discussed from many sides in Erice. My own take on this, based partially on Erice seminars runs roughly as follows:

(1) We are almost certain that the Earth is in a greenhouse, as suggested by Fourier in an article in Comptes Rendus in 1829 (I believe).

(2) We know from measurements starting with Tyndall in the 1870s that CO2 and H20 are greenhouse gases

(3) CO2 spreads around the world almost equally whereas, as I look out of my widow I notice that H20 is not spread uniformly

(3) We can calculate, as did Arrhenius in 1897, the temperature rise from a static doubling of CO2 (static in that we assume the doubling is sustained). IF water vapor stays the same.

(4) We know that as temperature goes up water evaporates. This is likely to accentuate the temperature rise due to CO2. We can calculate IF we assume that it is spread either uniformly (false) or a uniform relative increase.

(5) In Erice in 1979 Professor Kellogg presented the results of a number of models one of which had a negative feedback! Only recently have the models advanced enough to give a fair decryption of the past.

(6) In one small group it was mentioned that the feedback must be so large that the atmosphere "switch to a new region of stability". (whatever that means) Few scientific groups (such as the US National Academy of Sciences) are willing to put an error bar on their predictions, fearing that they would be misunderstood. Bolder persons such as myself try a Gaussian fit to the feedback factor and guess thereby that the probability of a large water level rise in New York City by the end of the century

is about the same as for a large nuclear power accident or about 1 in 10,000. As Professor Clarke said early in the 1980s - this should lead us to ask each other: "what is the meaning of such a statement?" Where else can this be discussed other than the Erice seminars on planetary emergencies?

(7) I go on. If I accept (reluctantly) that there will be a 2 degree world temperature rise by the end of the century, what is the effect on mankind? On average, on the poor, on the rich? This is even more uncertain.

(8) But I go on to be more philosophical and consider the purpose of these seminars. Does the huge uncertainty make you happy, so that you don't need to do anything, or unhappy, so that you constantly wonder what to do? I return to my thoughts noted above that these seminars on planetary emergencies should be on how to prevent accidental natural catastrophe. Uncertainty should lead us to seek information to reduce the uncertainty and unhappiness should lead us to think carefully about actions (bringing carbon out of the ground and emitting it as CO2) that force such considerations.

Even more uncertain and even more contentious have been the discussions about terrorism taking place mainly in the Permanent Monitoring Panel on Terrorism. We, like the rest of society, have been unable to come up with a useful definition of terrorism. Terrorist Acts are easier to define, and some progress has been made in discussion of their mitigation. I am not even sure whether there is agreement on how severe a problem it is. As I have noted (in Erice) that the 2,500 or so victims of the World Trade Center bombing pale into insignificance compared to the damage, both individual and societal, caused by the drug trade. I suspect that the number of Israelis killed annually by car accidents exceeds the number killed by terrorists. Yet it is the latter which dominates political actions. Again, what is the meaning of such a comparison and what should that imply for the future? Should society just ignore terrorism? I have never heard these properly discussed outside Erice.

In many of these examples the proper application of scientific knowledge and of scientific principles is hard to accomplish. The Erice seminars are only a small beginning. Occasionally there are situations which are (almost) scientific quarrels. But the general aim is there, led ably by Nino Zichichi. Let us all continue and improve upon his initiative.

Lowell Wood[1]

Then barely into my fifth decade of life, I came to the Erice "Seminars On Planetary Emergencies and the Prevention of Nuclear War" in the early '80s at the invitation of Professor Zichichi upon nomination of Professor Edward Teller.

During the previous decade, I had become a student and then a (characteristically, somewhat obstreperous) critic of much of the Conventional Wisdom within the U.S. Government regarding the nuclear weaponry aspects of national security. I once offered an (unwittingly, highly destructive) legalism-based criticism of the then-extant strategic nuclear war plans of the USA, and I followed that outrage by declaring that the then-highly classified PONAST (Post Nuclear Attack Survivability) studies made it clear that even in the comparatively 'mild' strategic nuclear war circumstances of that long-since-gone era, it was simply Pollyanna-ish in the extreme to pretend that the United States could continue as a nation in any modern sense of the word after an all-out nuclear exchange with the Soviet Union.

My punishment for these heresies was to be 'conscripted' in '74 as the junior-most member – by one if not two decades-of-age, as most all the other members were the famed "Wizards of Armageddon" – into the Senior Advisory Group on Effects (SAGE) of the then-Defense Nuclear Agency, a civilian body charged with advising the Defense Department on the technological aspects of waging nuclear war. [Mercifully enough, I was refused the rather unusual prerequisite security clearances – whereupon then-Vice President Rockefeller reportedly mandated their issuance to me; justice was to be served, no matter what!] Most of our time during my Group membership of several years was spent worrying every few months for days-on-end in well-upholstered, high-security dungeons not about "strategic" nuclear war, but rather as to how NATO was to wage a successful "tactical" nuclear war on the European continent against the invading forces of the Warsaw Pact. When I was finally allowed to 'retire' from this Group, the casualty rates in the highly-detailed NATO-Pact war-games were cresting at 1 division per hour annihilated on each side, as horrifying days melted into ghastly weeks – and my parting advice to my senior colleagues was to plainly recognize – and inform the Government – that this whole "tactical" nuclear war business was an adventure that no rational actor would ever commence, on either side, for any reason at all.

Quite persuaded by then that no one would ever want a nuclear explosive going off anywhere in their vicinity, I then looked quite intently into how they could be employed against each other, to insure that no one of them would never detonate at all close to people-&-valued things. These studies taught me again-&-again that there are an awful lot of pico-steradians – and even a lot of micro-steradians – in four-pi steradians, a pico-steradian and a micro-steradian being plausible bounds of the typical solid angle subtended by a defensive nuclear explosive at an offensive one, "out in the Real World." The "Bottom Line" of years of supercomputer-aided studies was that even <u>very</u> high-yield defensive systems didn't put all that much potentially-negating energy on plausibly-distant offensive ones; furthermore, these offensive ones could be made to be remarkably durable against such defensive actions. Net, use of (even rather extra-)ordinary nuclear explosives to defend against other such explosive seemed "quantitatively challenged" – quite innately so. Teller found these results at least as dismaying as did I – but "facts are stubborn things."

[1] Research Fellow, Hoover Institution, Stanford University, Stanford CA 94305 USA. Email: lowellwood@comcast.net

Interest naturally turned to means of brightening, or focusing, energy from such defensive systems, by one at-least-arguably practical means-or-another. Lasers of various types innately appealed, the more so as some of these, e.g., ones operating in the x-ray portion of the electromagnetic spectrum, were intrinsically very mass-efficient, as well as being mechanistically well-matched to exceedingly high energy-density "power supplies" such as nuclear explosives. Other types of lasers – indeed, of directed energy systems of several different kinds – were also of very considerable interest. All of this 'fit' within the charge that I'd been given at the very outset of the Reagan Administration to bring all such high-tech systems[2] forward as swiftly as ever possible, as part-&-parcel of Reagan's long-before-announced but universally-ignored US grand strategy for confronting the Soviet empire: "It's really very simple: we win; they lose."

One of the most memorable Erice Seminars, in August 1983, took place against the freshly-announced Strategic Defense Initiative of the Reagan Administration. [To be sure, I had extensively supported Teller's advocacy of this Initiative within the private councils of the Administration in the preceding two years – and had watched as the Joint Chiefs of Staff had given their unanimous endorsement to it, well before Reagan's speech announcing it – so it didn't come as that much of a surprise to my colleagues and myself, compared to what many others found it to be.]

Having very freshly observed the "orderly chaos" within the US national security world as it struggled to cope with the newly-announced (albeit hardly new-in-actuality) Presidential policy – commencing with figuring-out just what it was, let alone how to implement it, and what its side-effects and consequences might be – I was struck by the dismay-bordering-on-consternation of the individual-&-collective reaction of the delegates from the Soviet Union. They clearly had very little idea – even less than did several of the Americans present – of what "strategic defense" was, but they knew (quite positively!) that they were adamantly opposed to it. Teller-&-my assurances to the effect that it was merely an expression of what Soviet Premier Alexei Kosygin had eloquently advocated just prior to the Glassboro Summit in 1967 – "the moral duty of any government to defend its people against attack" – weren't positively received by the Soviet representatives present (to phrase the matter delicately).

We Americans were well aware that the Soviet government's nascent position had developed against the background of intensive, not-so-clearly-decided debate within senior advisory circles in the USSR as to whether "strategic defense" was a 'Good Idea' or not, especially given the continuing huge investments in urban-scale defenses against strategic rocket-&-bomber attacks which the Soviet government had been making since the '60s. We therefore pressed our Soviet interlocutors vigorously and persistently on all aspects of the set of issues, ranging from technical feasibility to philosophical aspects of modern strategic warfare – and they responded in an entirely predictable, indeed highly-stylized manner. These debates were the more "free-wheeling" as none of us were representing our respective governments and each of us were free to present our individual viewpoints – so that our Soviet colleagues seemed forever-puzzled as to exactly which one was that of the American government on any

[2] In retrospect, of course, we culminated the Strategic Defense Initiative with defensive systems centered on focused-&-directed "beams" of matter, not energy: a set of remarkably compact and high-performance satellites in low Earth orbit, each capable of fully autonomous, long-duration operation in a military environment, which could be operated to efficiently place themselves precisely in the paths of attacking ballistic missiles – with the resulting hypervelocity collision inducing the complete destruction of both the defensive satellite and the attacking missile. Some who didn't understand – or believe – the underlying mandate to be exceptionally technologically opportunistic in the defeat of long-range missiles with non-nuclear means were puzzled by our seeming indifference to choice-of-defensive means.

point-in-contention – if indeed any of them were! The much more monolithic Soviet set-of-viewpoints thus was peppered with critiques from many distinct perspectives – which of course complicated Soviet responses to them.

It was feasible to have such sustained, intensive discussions in the early '80s – in an era in which the Nuclear Freeze movement was at its peak in the NATO nations, when Reagan and Andropov were leading the two superpowers and when the two superpower nuclear war machines were at their most "bristle-y"[3] – only because of the stern even-handedness and exquisite, exceedingly-perceptive attention to detail with which Nino Zichichi structured and then operated these Seminars. It was completely evident to all onlookers that the two superpowers' viewpoints were being advocated by exceptionally strong personalities -- and the Europeans and Asians presently certainly weren't limp-wristed with respect to their various perspectives. This was a situation "that could not long endure," but Professor Zichichi nonetheless "made it so" – aided by a most adept-&-perceptive Majorana Center staff led by Maria Zaini and Claude Manoli.

There were many essential features which contributed to this extraordinary success, but some of them stand-out in memory particularly vividly, including Nino's exquisitely-crafted – and highly dynamically 'amended' – schedules-of-topics-&-speakers, his implacable demands for speakers' adherence to assigned speaking-times-&-topics, and the iron resolve so surefootedly and exquisitely blended with classic European politeness with which he 'refereed' the essential but often highly-contentious exchanges and 'interventions' after presentations. [Indeed, if conducted under American-style "rules of engagement", the Seminars surely would have centripetally disassembled after a single day – but Nino kept them going most productively, year-after-year!]

An utterly-crucial aspect of the Seminars – which I didn't realize clearly except in retrospect – were the grand dinners that Nino convened each evening for all Seminar participants, and over which he most carefully-&-vigorously presided – especially the portion conducted at "Professor Zichichi's table," at which he personally seated his selected guests from the day's fray and the delegation-leads, even to the point of assigning seating relative to his own, so that he could 'intermediate' vigorously – and most effectively. Looking back on these every-evening banquets (at which attendance was *de facto* obligatory), I now realize that these were probably crucial to keeping the Seminars "from running off the rails," for it's difficult for most any civilized person to wine-&-dine-&-converse with someone the previous evening for a few hours and then credibly work-up a ruinous fury with them the following morning – or even afternoon -- and Nino seemed to masterfully exploit this. With many such means wielded with High Sicilian artistry, he kept not only each such Seminar working productively, but maintained the operation of the entire Seminar series in a highly-effective fashion. *Mirabile visu!*

I remembered in detail these Seminars – which had a quite gripping character over a dozen-year interval culminating in the dissolution of the Soviet Union – but I really didn't ponder why-&-how they could have done what they did until very recently, when I've have the

---

[3] 'Twas the geopolitical season in which the Soviet SS-20 and USA Pershing II "theater" nuclear-armed missiles were being intensively deployed in Europe, KAL 007 was shot down in the Far East and all superpower arms control negotiations were suspended, an era culminating in the April '84 strategic nuclear war "surprise exercise" involving representative components of the entire Soviet nuclear war machine being synchronously "put through their paces" – thereby inducing a degree of 'sobriety' in Western security thinking-&-planning probably not seen since the Cuban Missile Crisis.

opportunity to study a number of hitherto-sequestered documents of the Reagan Administration, including minutes from key meetings of the US National Security Council and Reagan's own daily-diary, correspondence and manuscripts on strategic policy issues, especially those relating to nuclear war.[4]

These vividly etch the USA national policy background against which the Seminars were conducted – and they hint broadly at detailed mechanistic matters likely known only to Professor Zichichi among those still with us. Of particular likely relevance in these latter respects are the just-now-published, remarkable 'early' policy correspondence between President Reagan and Pope John Paul II, President Reagan's receiving of 'purposeful' papal emissaries (particularly the Vatican State Secretary Cardinal A. Casaroli) for 'closed discussions' – as well as the still-sealed-in-the-USA 'later' personal correspondence between Reagan and the Pope. These dovetail to highly suggestive – if not downright compelling -- extents with the existence and content of the papal greetings-&-blessings which Professor Zichichi invariably opened each year's Seminar – and which 'naturally' elicited comparable greetings-&-best wishes from the heads of state of Italy, the USSR and the USA.[5]
The principal purpose of the Seminars of a historic nature is likely the opportunities afforded for senior technical advisors of the Soviet and American governments to exchange viewpoints in quite candid manners – reasonably candid during Seminar sessions themselves "while the whole world was watching," and sometimes-brutally-candid exchanges during off-line discussions during breaks, meal-times, etc. – ones not infrequently mediated by Nino himself. I can testify that senior American attendees "reported back" on Seminar events to senior levels of their government, often with telling impacts (e.g., on the course of the Strategic Defense Initiative, which was quite 'responsive' – by design – to Soviet reactions-&-perceptions). I have reason to believe that the situation was basically symmetric on the Soviet side.[6]

---

[4] Those interested in the history of this epoch will do well to carefully study "Reagan's Secret War: The Untold Story Of His Fight To Save The World From Nuclear Disaster," by Martin Anderson and Annelise Anderson, being published in June 2009 by Crown Publishers of New York City; ISBN 978-0-307-23861-0. This work, introduced by Reagan's Secretary of State George Shultz and replete with presentation-&-authoritative analyses of a plethora of hitherto-classified policy and strategy documents of the Reagan presidency, seems likely to be foundational with respect to the strategic policy-related history of this era.

[5] The general nature of this 'compulsion' was made vividly clear to me one year when the Seminar actually opened without any such greetings. Teller conferred privately in an increasingly agitated fashion with Zichichi at the day's beginning and during each break on this matter, and required me to 'poke at' the DC System – then in the midst of its August vacation, as well as five hours earlier-in-the-day – by telephone after most such occasions; I did so, but quite ineffectively re "results". Finally, in late-PM, he demanded that I accompany him to Zichichi's personal office in San Rocco, where he picked up the Institute's only "overseas call-capable" phone – and broke out in a sustained stream of Italian! A few minutes later of first-Italian-then-English-then-Italian-then…, he was speaking in clipped, formal English to some senior staffer in the White House in Washington DC, demanding in a tightly-controlled fury that the President's official greetings to the Seminar be dispatched forthwith – lest he have to take the matter to the President personally. [I expressed surprise at the end of this phone-call that he "spoke Italian", to which he unhappily replied, "I read Italian, but I really don't speak it. That was simply 'telephone Italian,' which even I can manage!" Indeed, several years previously, as he and I were air-traveling to an early Seminar, I found him late-at-night over the Atlantic reading Dante's *Divine Comedy* in Italian; to my expression of surprise that Italian was among the half-dozen 'foreign' languages at his command, he replied that he was merely "brushing up a bit". I suggested that Dante-in-the-original was rather archaic Italian, to which he responded that the pleasantness of this particular language-lesson traded off well against its questionable utility.] The demanded Presidential greetings arrived barely an hour later – and Professor Zichichi imperturbably presented all four of the heads-of-state greetings to the Seminar at the conclusion of that first day's Seminar session – as though this were the exact-&-customary manner in which such greetings were to be conveyed.

[6] The "human aspects" of the Soviet-American interactions at the Seminars is perhaps illustrated by an incident that occurred in Summer '92, when Teller made his first-ever trip to Russia, and was being feted, right off the

A secondary purpose of the Seminars of historical impact was to enable senior Soviet and American representatives – often including those holding official or declared positions – to exchange views and even conduct negotiations. This occurred substantially more often during the Seminars' latter interval, as the Gorbachevian reforms made it feasible for such interactions to occur and gave them the possibility of being productive. Nino was an exceptionally gracious and sensitive host in such matters, even to the extent of making his own private Institute Director's office available for one such set of negotiations of great possible moment.[7]

Two incidents stand-out in memory especially vividly from the Seminars.

In the first, Paul Dirac – in whose honor the great hall of the Majorana Center is named and whose Equation so splendidly graces its presidium in replica-manuscript – spoke in characteristically dry-&-detached but intensely moving fashion of his professional collaboration and personal friendship of six decades with Pyotr Kapitza, who sat in the front row during this talk. The loyalty-&-heroism of this most unusual man[8] in repeatedly

---

plane, at the Igor Kurchatov villa in Moscow, complete with hordes of media representatives wielding video cameras. Spotting Mme. Kurchatova's grand piano in the formal parlor of the exceptionally well-conserved villa which Stalin had personally commissioned to reward Kurchatov for his exceptional services to USSR military-nuclear capabilities (Kurchatov had died there 32 years previously), Teller – then 84 and severely jet-lagged from no-stopover air-travel from San Francisco – leapt to its bench without saying a word and commenced to pound-out a remarkably credible start-to-finish rendering of the piano portion of Beethoven's _Archduke Trio_ (a real favorite of his, as well as my own) with characteristic care-&-passion. As he concluded this "artistic victory lap" in the very parlor of his Soviet opposite-number most of an hour later, a senior Soviet representative at the Seminars whispered to me, "I now better understand why you won and we lost." To my subsequent, privately-conveyed, "Not bad at all, considering your age-&-present condition," Teller's crisp riposte was, "And considering the piano's incredibly-poor maintenance – even you must have noticed how out-of-tune it was!"

[7] Occurring during the Summer '92 Seminar, this was quite possibly the last substantive contact at the top level between the now-Russian and American governments on the prospect of shared defenses against strategic attacks, which had been agreed-to-in-principle between Presidents Yeltsin and Bush during the Washington Summit of January '92. As I sat watching from the office's corner (invited to be there in order to provide military-technical commentary, if-&-as indicated), the American and Russian seniors negotiated back-&-forth, but were unable to conclude the sought-after agreement. As the three of us walked back to our respective quarters in San Rocco in a necessarily somber mood, I ventured that this might be last opportunity in a long while to conclude such an agreement, to which the senior Russian candidly responded, "We believe that we'll get a better offer from the next Administration!" – to which I responded, "Perhaps, but perhaps not. What then?" In fact, the next Administration declined to make any such offer, although it did cancel the space-based missile-interceptor program then – and still! – of greatest concern to the Russians. However, unresolved strategic defense issues continue to "poison the well" of Russian-American relations through the present day. Alas!

[8] Surely most-if-not-all physicists-in-training learn of the extraordinary diffidence, shyness and aloofness of the younger Dirac. My favorite anecdote along these lines was related by Teller, who averred that he heard it directly from the Dirac house-guest involved. One day, this person called upon Dirac at his home, and in the course of an extended discussion the two of them were served refreshments by a striking young lady. Not having been introduced, he couldn't refrain from asking Dirac as he was departing, "And who was the young lady?" "Oh," said Dirac with characteristic diffidence, "that was Wigner's sister." The statement was literally true – it was indeed the sister of Eugene Wigner, a close friend of Dirac's – but she was also Dirac's bride of several months! [To be sure, slightly different versions of this anecdote exist.]

A lesser-known side of Dirac – one that better illustrates just how steely he was capable of being in Kapitza's defense in the same season-of-life – was related to me in a long-ago, non-public discussion by Richard Feynman, who recalled criticisms of his work in quantum electrodynamics by Dirac and Teller at a closed physics conference at Cornell in the later '40s as being the next-most-severe to which he'd ever been subjected (the first being an ever-earlier post-seminar flogging at Princeton to which he'd been subjected by Pauli, albeit with concise-&-pungent 'rescue' immediately extended by Einstein). In Feynman's telling, Dirac had excoriated – with uncharacteristic passion – Feynman's presentation of the diagram methodology for which he'll be forever-remembered as "cartoon physics", and Teller had followed immediately by sharply criticizing its then-near-total

throwing his body in front of Kapitza's while the latter was "on death row" in Stalin-&-Beria's Soviet Union has long since been legendary in the annals of physics, but hearing him – then at death's door himself – speak of it extemporaneously was absolutely riveting.

The second incident played out over the entire '91 Seminar. Just before that year's Seminar convened, a group of reactionary Soviet political and military figures staged a coup d'etat against the Gorbachev government during the traditional "August vacation interval" so beloved of modern Europeans. As a consequence, the large group of senior Russians present at the Seminar found themselves in understandably-agonizing doubt as to the character of their country's government – and thus of their nation's future. The Majorana Center's able staff rigged-up an antenna just-barely-capable of receiving Moscow TV satcom broadcasts intended for reception much further east, and most Russians – and many of us non-Russians – were 'glued' to the TV monitors in the San Rocco computer-area during all-too-much of the Seminar-week. There we watched the coup falter-&-fail and the Gorbachev government restored to power in a bloodless revolution, highlighted by the days-long confrontation before the Russian Parliament building between the putschist tanks and a pro-democracy crowd of Moscovites famously rallied by Yeltsin's historic speech-from-atop-a-tank. In the course of this historic week,[9] one of the most senior members of the Soviet representation was

---

lack of formal basis. [In an informal memorial session for Feynman two decades ago, I related this incident and Feynman's remark that each of these two events had very nearly driven him from physics "forever," to which Teller responded, "I'm very sorry that you didn't tell me of this before Dick died. I remember well the occasion, and I should have apologized to him long since for the harshness of what I then said, especially since Freeman (Dyson) fixed the problem so well and so soon thereafter – but I had no idea until just now that he took it so badly."]

[9] On the Seminar's concluding day, I was tasked with moderating the opening session, and I abused the discretion extended to me by first hailing-from-the-chair the triumph of the forces of freedom with a mini-speech scribbled over breakfast. Said I, in this intervention entitled "All Glory To The Heroes Of The August Revolution:"

"It is a genuine, high honor to address an international conference which includes such extensive representation from the Soviet Union at the end of a historic week which saw the citizens of this, the world's third most populous nation, rise up *en masse* against their would-be oppressors and successfully demand freedom and democracy. As is often the case --- any where and in any era -- the revolutionary vanguard of the Moscow Uprising rose up at the immediate risk of their lives, for -- as Lenin so aptly warned -- the oppressors of the people will never yield power voluntarily, but will have to have it ***wrested*** from them by the people.

"I will frankly confess to you that I, as well as many other of my fellow Americans, have long feared that Mother Earth would be far more blood-soaked --- and perhaps more radioactive, as well --- before the statue of Felix Dzerzhinsky, the arch-fiend who killed more Russians during the Lenin Terror than had all the Polish armies in history, would be toppled from its pedestal in Moscow.

"Yesterday, however, we watched on international TV as the heroes of the August Revolution smashed the toppled Dzerzhinsky statue into rubble with hand tools.

"Yesterday, we heard the first freely elected leader of the Russian people in a thousand years ban all activities of the Communist Party in the territory of the Russian Federation and then suspend indefinitely the publication of that "Mother of All Lies," the Communist Party newspaper *Pravda.*

"Yesterday, we heard the President of the Soviet Union and the General Secretary of the Communist Party of the Soviet Union publicly and sharply criticize the Party's Central Committee, the body which made him what he is today.

"Truly, we live in revolutionary times, but strange ones which bode well for all mankind. A half-century ago, Winston Churchill addressed the people of newly Nazi-conquered Europe, commanding them to continue to resist the fascists and promising that they would ultimately triumph whereupon, he prophesied, all mankind would attain to "the broad uplands, where the sun will shine brightly upon the brave and the true and gloriously on the tombs of heroes.'

"These broad uplands now seem very near at hand indeed, for the last fascists on the European continent have finally been beaten down by the most brave-and-true of the heroic Russian people. To them, to our Slavic brethren, belongs the re-doubled glory of breaking the back of the German National Socialist tyranny a half-century ago and, just this week, that of its Russian cousin.

summoned to return most urgently to Moscow, in the same basic manner that had eventuated so personally catastrophically for so many of his countrymen in decades past, i.e., to an entirely uncertain fate. I happened to be around a hallway corner in the San Rocco reception area in the pre-dawn hour when we were both about to embark from the Majorana Center for the Palermo airport and flights toward our respective homelands – and I inadvertently overheard him appeal in the most stark human terms to Nino for sanctuary-&-succor for those of his family who had accompanied him to the Seminar, which plea was immediately and very graciously granted. A most intensely-memorable week – and a most excellent human being!

The Seminars On The Prevention Of Nuclear War are now history, some small-but-crucial portions of which they helped write. All of us privileged to live in the post-Cold War era owe a large debt-of-gratitude to Prof. Nino Zichichi for his vision, daring, initiative and many years of indefatigable endeavors to bring the Seminars into existence and then to see them bear their bountiful fruit, for our present safety-&-serenity on international scales surely are due in no small part to him and his Seminar efforts!

"Though Russians have arrived late in the community of free men, they come in the splendor and with the honor of having defeated both of the two worst tyrannies of modern times; they have proved beyond any possible doubt that they know how to die in defense of their homeland and their freedoms.

"One final momentous task stands before the freedom-loving peoples of the Earth, for, as an American patriot warned two centuries ago, "When human freedom is threatened anywhere, it is in danger everywhere." The ancient totalitarians which still oppress the great Chinese people continue to defy the verdict of history already rendered in every other major nation, their hands red with the blood of the Chinese people while already their own children curse them and deny their parentage.

"Let us hope and pray that, by the time we next meet here, the Chinese people will have fought successfully for their freedom, that our Chinese colleagues and particularly our friend, the Chinese Ambassador, will be here as the representatives of a democratically governed Chinese nation, and that the butchers of Tiananmen Square will have joined Lenin, Stalin, Hitler and Mao, the other great monsters in human guise who have sorely afflicted humankind in the 20th century, in whatever Circle of the Inferno is reserved for them.

"This personal note concluded, now please let me speak to the subject at hand, some aspects of the ways-and-means of implementing the Global {Environmental] Monitoring Program."

PHOTO GALLERY

Cover photo: Erice as a scientific summit (Photo courtesy of W. A. Barletta)

Erice Mayor Ignazio Sanges with Professor Antonino Zichichi, preparing to welcome the participants to the Erice International Seminars in 2005 (Photo courtesy of W. A. Barletta)

The site of the temple of Venus at Erice, as celebrated by Virgil in the Aeneid, later became the site of fortress castles after the Norman invasion. The mystically evocative powers of Erice are described in the testimonials of W. Barletta and V. Krivokhizha (Photo courtesy of W. A. Barletta)

The beach at San Vito lo Capo was a frequent choice for an excursion by participants in early years following the closing Plenary Session. In late August the beach would be teaming with vacationing Italians (Photo courtesy of W. A. Barletta)

The panoramic view from the break room in San Domenico (now the P.M.S. Blackett Institute) described in the testimonial of J. Westby (Photo courtesy of W. A. Barletta)

Professor Robert Clark (left) and his colleagues at a coffee break during the plenary sessions of 2004. The setting provoked both deep conversation and introspection (Photo courtesy of W. A. Barletta)

In the front row all together for the last time (2002), Seminar Chairman Antonino Zichichi with Seminar co-chairs T. D. Lee (left) and Kai M. B. Siegbahn (right)

Nobel Laureate Paul A. M. Dirac at the first International Seminar, one of the authors of the Erice Statement, together with Piotr Kapitza and Antonino Zichichi

Nobel Laureate I.I. Rabi at Erice

1979 Nobel Laureate Abdus Salam at Erice

Nobel Laureate Samuel C. C. Ting has been a frequent participant at Erice

Edward Teller confers with Paul Dirac during a coffee break at the First International Seminar

East meets West: As recalled by B. Galland and W. Barletta, the encounters between Edward Teller (left) and Evgeny Velikhov (right) were a hallmark of the first ten years of International Seminars (Photo courtesy of B. Galland)

Science Advisors - East meets West on the Erice podium. Clasping hands presidential science advisors Qian Jiadong China, Edward Teller USA and Evgeny Velikhov USSR, together with Antonino Zichichi (idem photo below)

Academician Evgeny Velikhov and Italian Foreign Minister Giulio Andeotti and Antonino Zichichi on a visit to Russia, hosted by the USSR Minister of Foreign Affairs, Edward Shevardnadze

Juras Pozela (left), President of the Lithuanian Academy of Sciences, Evgeny Velikhov (lower right) and Karl Rebane (upper right), President of Estonian Academy of Sciences in the Paul A. M. Dirac lecture hall in 1992 (Photo courtesy of J. Pozela)

Press conference in March 1992: media representatives (leftmost two) with Juras Pozela, Kai Siegbahn, Antonino Zichichi and Evgeny Velikhov (Photo courtesy of J. Pozela)

Edward Teller's closest collaborator, Lowell Wood, was a frequent participant during the early years as recalled by Bertil Galland (Photo courtesy of B. Galland)

In 2002 Edward Teller's last address to the International Seminars was via a teleconference link

a) 

b) 

Recollections of a focus on Information Security in 2008

a) Prof. Zichichi with International Telecommunications Union Secretary General, H. Touré (left) and Dr. Robert Khan (right)

b) A later encounter at the Erice Prize Award Session: Dr. Touré (left), Mrs. Touré (center) and Information Security PMP member V. Britkov (Photos courtesy of V. Britkov)

Lunch at Erice in 1999 with Juras Pozela (left), Zenonas Rudzikas (center) and Albert Tavkhelidze (right), President of the Georgian Academy of Sciences

Presentation of the Kapitza Gold Medal to Chancellor Lorne Everett at the World Federation of Scientists meetings on October 29th, 1999 at the Beau Rivage Palace in Lausanne Switzerland. Medal presentation by Dr Igor S. Zektser on behalf of the Russian Academy of Sciences. L to R: Dr Antonino Zichichi, Chancellor Lorne G. Everett, Dr Igor Zektser, Dr Kai M B. Siegbahn, Nobel Laureate, Dr Tsung-Dao Lee, Nobel Laureate and Dr Gerardus't Hooft, Nobel Laureate

YEARBOOKS

THE SCIENCE AND CULTURE SERIES
Nuclear Strategy and Peace Technology

Series Editor: Antonino Zichichi

1981 — International Seminar on Nuclear War — 1st Session: The World-wide Implications of Nuclear War

1982 — International Seminar on Nuclear War — 2nd Session: How to Avoid a Nuclear War

1983 — International Seminar on Nuclear War — 3rd Session: The Technical Basis for Peace

1984 — International Seminar on Nuclear War — 4th Session: The Nuclear Winter and the New Defence Systems: Problems and Perspectives

1985 — International Seminar on Nuclear War — 5th Session: SDI, Computer Simulation, New Proposals to Stop the Arms Race

1986 — International Seminar on Nuclear War — 6th Session: International Cooperation: The Alternatives

1987 — International Seminar on Nuclear War — 7th Session: The Great Projects for Scientific Collaboration East-West-North-South

1988 — International Seminar on Nuclear War — 8th Session: The New Threats: Space and Chemical Weapons — What Can be Done with the Retired I.N.F. Missiles-Laser Technology

1989 — International Seminar on Nuclear War — 9th Session: The New Emergencies

1990 — International Seminar on Nuclear War — 10th Session: The New Role of Science

1991 — International Seminar on Nuclear War — 11th Session: Planetary Emergencies

1991 — International Seminar on Nuclear War — 12th Session: Science Confronted with War (unpublished)

1991 — International Seminar on Nuclear War and Planetary Emergencies — 13th Session: Satellite Monitoring of the Global Environment (unpublished)

1992 — International Seminar on Nuclear War and Planetary Emergencies — 14th Session: Innovative Technologies for Cleaning the Environment

1992 — International Seminar on Nuclear War and Planetary Emergencies — 15th Session (1st Seminar after Rio): Science and Technology to Save the Earth (unpublished)

1992 — International Seminar on Nuclear War and Planetary Emergencies — 16th Session (2nd Seminar after Rio): Proliferation of Weapons for Mass Destruction and Cooperation on Defence Systems

1993 — International Seminar on Planetary Emergencies — 17th Workshop: The Collision of an Asteroid or Comet with the Earth (unpublished)

1993 — International Seminar on Nuclear War and Planetary Emergencies — 18th Session (4th Seminar after Rio): Global Stability Through Disarmament

1994 — International Seminar on Nuclear War and Planetary Emergencies — 19th Session (5th Seminar after Rio): Science after the Cold War

1995 — International Seminar on Nuclear War and Planetary Emergencies — 20th Session (6th Seminar after Rio): The Role of Science in the Third Millennium

1996 — International Seminar on Nuclear War and Planetary Emergencies — 21st Session (7th Seminar after Rio): New Epidemics, Second Cold War, Decommissioning, Terrorism and Proliferation

1997 — International Seminar on Nuclear War and Planetary Emergencies — 22nd Session (8th Seminar after Rio): Nuclear Submarine Decontamination, Chemical Stockpiled Weapons, New Epidemics, Cloning of Genes, New Military Threats, Global Planetary Changes, Cosmic Objects & Energy

1998 — International Seminar on Nuclear War and Planetary Emergencies — 23rd Session (9th Seminar after Rio): Medicine & Biotechnologies, Proliferation & Weapons of Mass Destruction, Climatology & El Nino, Desertification, Defence Against Cosmic Objects, Water & Pollution, Food, Energy, Limits of Development, The Role of Permanent Monitoring Panels

1999 — International Seminar on Nuclear War and Planetary Emergencies — 24th Session: HIV/AIDS Vaccine Needs, Biotechnology, Neuropathologies, Development Sustainability — Focus Africa, Climate and Weather Predictions, Energy, Water, Weapons of Mass Destruction, The Role of Permanent Monitoring Panels, HIV Think Tank Workshop, Fertility Problems Workshop

2000 — International Seminar on Nuclear War and Planetary Emergencies — 25th Session: Water — Pollution, Biotechnology — Transgenic Plant Vaccine, Energy, Black Sea Pollution, Aids — Mother–Infant HIV Transmission, Transmissible Spongiform Encephalopathy, Limits of Development — Megacities, Missile Proliferation and Defense, Information Security, Cosmic Objects, Desertification, Carbon Sequestration and Sustainability, Climatic Changes, Global Monitoring of Planet, Mathematics and Democracy, Science and Journalism, Permanent Monitoring Panel Reports, Water for Megacities Workshop, Black Sea Workshop, Transgenic Plants Workshop, Research Resources Workshop, Mother–Infant HIV Transmission Workshop, Sequestration and Desertification Workshop, Focus Africa Workshop

2001 — International Seminar on Nuclear War and Planetary Emergencies — 26th Session: AIDS and Infectious Diseases — Medication or Vaccination for Developing Countries; Missile Proliferation and Defense; Tchernobyl — Mathematics and Democracy; Transmissible Spongiform Encephalopathy; Floods and Extreme Weather Events — Coastal Zone Problems; Science and Technology for Developing Countries; Water — Transboundary Water Conflicts; Climatic Changes — Global Monitoring of the Planet; Information Security; Pollution in the Caspian Sea; Permanent Monitoring Panels Reports; Transmissible Spongiform Encephalopathy Workshop; AIDS and Infectious Diseases Workshop; Pollution Workshop

2002 — International Seminar on Nuclear War and Planetary Emergencies — 27th Session: Society and Structures: Historical Perspectives — Culture and Ideology; National and Regional Geopolitical Issues; Globalization — Economy and Culture; Human Rights — Freedom and Democracy Debate; Confrontations and Countermeasures: Present and Future Confrontations; Psychology of Terrorism; Defensive Countermeasures; Preventive Countermeasures; General Debate; Science and Technology: Emergencies; Pollution, Climate — Greenhouse Effect; Desertification, Water Pollution, Algal Bloom; Brain and Behaviour Diseases; The Cultural Emergency: General Debate and Conclusions; Permanent Monitoring Panel Reports; Information Security Workshop; Kangaroo Mother's Care Workshop; Brain and Behaviour Diseases Workshop

2003 — International Seminar on Nuclear War and Planetary Emergencies — 29th Session: Society and Structures: Culture and Ideology — Equity — Territorial and Economics — Psychology — Tools and Countermeasures — Worldwide Stability — Risk Analysis for Terrorism — The Asymmetric Threat — America's New "Exceptionalism" — Militant Islamist Groups Motives and Mindsets — Analysing the New Approach — The Psychology of Crowds — Cultural Relativism — Economic and Socio-economic Causes and Consequences — The Problems of American Foreign Policy — Understanding Biological Risk Chemical Threats and Responses — Bioterrorism — Nuclear Survivial Criticalities — Responding to the Threats — National Security and Scientific Openness — Working Groups Reports and Recommendations

2003 — International Seminar on Nuclear War and Planetary Emergencies — 30th Session: Anniversary Celebrations: The Pontifical Academy of Sciences 400th — The 'Ettore Majorana' Foundation and Centre for Scientific Culture 40th — H.H. John Paul II Apostolate 25th — Climate/Global Warming: The Cosmic Ray Effect; Effects on Species and Biodiversity; Human Effects; Paleoclimate Implications; Evidence for Global Warming — Pollution: Endocrine Disrupting Chemicals; Hazardous Material; Legacy Wastes and Radioactive Waste Management in USA, Europe; Southeast Asia and Japan — The Cultural Planetary Emergency: Role of the Media; Intolerance; Terrorism; Iraqi Perspective; Open Forum Debate — AIDS and Infectious Diseases: Ethics in Medicine; AIDS Vaccine Strategies — Water: Water Conflicts in the Middle East — Energy: Developing Countries; Mitigation of Greenhouse Warming — Permanent Monitoring Panels Reports — Workshops: Long-Term Stewardship of Hazardous Material; AIDS Vaccine Strategies and Ethics

2004 — International Seminar on Nuclear War and Planetary Emergencies — 31st Session: Multidisciplinary Global Approach of Governments and International Structures: Societal Response — Scientific Contributions to Policy — Economics — Human Rights — Communication — Conflict Resolution — Cross-Disciplinary Responses to CBRN Threats: Chemical and Biological Terrorism — Co-Operation Between Russia and the West — Asymmetrical Conflicts — CBW Impact — Cross-Disciplinary Challenges to Emergnecy Management, Media Information and Communication: Role of Media in Global Emergencies — Emergency Responders — Working Groups' Reports and Recommendations

2004 — International Seminar on Nuclear War and Planetary Emergencies — 32nd Session: Limits of Development: Migration and Cyberspace; in Europe; Synoptic European Overview; From and Within Asia; Globalization — Climate: Global Warming; a Chronology; Simple Climate Models; Energy and Electricity Considerations — T. S. E.: CJD and Blood Transfusion; BSE in North America; Gerstmann-Straussler-Scheinker Disease — The Cultural Emergency: Innovations in Communications and IT — Cosmic Objects: Impact Hazard; Close Approaches; Asteroid Deflection; Risk Assessment and Hazard Reduction; Hayabusa and Follow Up — Aids and Infectious Diseases: Ethics in Medicine; International Co-operation; Laboratory Biosecurity Guidelines; Georgian Legislation; Biosecurity Norms and International Organizations, Legal Measures Against Biocrimes — Water and Pollution: Cycle Overview; Beyond Cost and Price; Requirements in Rural Iran; Isotope Techniques; Clean and Reliable Water for the 21st Century — Permanent Monitoring Panels Reports — Workshops: Global Biosecurity; Cosmic Objects

2005 — International Seminar on Nuclear War and Planetary Emergencies — 34th Session: Energy: Nuclear and Renewable Energy; Energy Technologies for the 21st Century; Repositories Development; Nuclear Power in Europe and in Asia; The Future of Nuclear Fusion — Climate: Global Warming; Celestial Climate Driver; Natural and Anthropogenic Contributions; Climate Data and Comparison with Models; Understanding Common Climate Claims — AIDS and Infectious Diseases: New Threats from Infectious Agents — SARS Epidemic; Vaccines Development; Transmissible Spongiform Encephalopathies Update — Limits of Development: International Points of View on Migration — Pollution: Science and Technology; Subsurface Laser Drilling — Desertification: A Global Perspective; Integrated Approaches — Disarmament and Cultural Emergencies: A WFS Achievement in China; Non-Proliferation — Permanent Monitoring Panel Reports — Workshops: Energy; Information Security; Building Resilence Associated with the Third Meeting on Terrorism

2006 — International Seminar on Nuclear War and Planetary Emergencies — 36th Session: Energy: Global Nuclear Power Future; Global Monitoring of the Planet Proliferation: Nuclear Weapons; AIDS and Infectious Diseases: Avian Flu — Global Health; Climatology: Global Warming/Aerosols and Satellites; Pollution: Plastic Contaminants in Water; Information Security: Relevance of Cyber Security; Limits of Development: Development of Sustainability; Defence Against Cosmic Objects; WFS General Meeting: Cultural Energy-Focus: Terrorism; Permanent Monitoring Panel Reports; Limits of Development Permanent Monitoring Panel Meeting; World Energy Monitoring Workshop.

2007 — International Seminar on Nuclear War and Planetary Emergencies — 38th Session: World Energy Crisis; Managing Climate Change; Mitigation of Greenhouse Gases; Geoengineering & Adaptation; Theoretical Alternatives to Climate Modelling; US Missile Defence Shield; Global Monitoring of the Planet; Life Cycle Nuclear Energy Environmental Issues; The Epidemic of Alzheimer; Infectious Agents and Cancer

2008 — Energy: Nuclear Power Present and Future; Sustainability of Biofuels; Resolving the Nuclear Waste – Climatology Model and Statistics; Ozone and Climate Change Interaction; Spatio-Temporal Field of Atmospheric CO_2; Forest Policies – Medicine: Vector-Borne Diseases; Screening Technology – Pollution: Air-Borne Particulates – Global Monitoring of the Planet: Disarmament and Non-Proliferation Regime; The Crisis in Internet Security; The Northern Sea Route – The Erice Science for Peace Award Scientific Session

PERMANENT MONITORING PANELS

Permanent Monitoring Panels

The identification of Planetary Emergencies and their categorisation into fifteen main groups, with 56 subgroups, was one of the most important early achievements of the World Federation of Scientists. Today, it leads the fight against the Planetary Emergencies, a field of endeavour of vital importance for both the Developed and the Developing Countries throughout the world, and one in which a united, international effort will have greater impact than the achievements of individual or groups of countries.

To date the World Federation of Scientists has established the following Permanent Monitoring Panels and Working Groups:

- Biotechnology
- Brain and Behaviour
- Climatology
- Defence Against Cosmic Objects
- Desertification
- Energy
- Floods and Extreme Weather Events
- Information Security
- Limits of Development
- Missile Proliferation
- Mother & Child
- Pollution
- Motivations for Terrorism
- Mitigation of Terrorist Acts
- Transmissible Spongiform Encephalopathy
- Water

CONTRIBUTORS

Professor Andrei Andries
Honorary President
The Academy of Sciences of Moldova
WFS National Scholarship Programme
Chisinau, Moldova

Professor William A. Barletta
U.S. Particle Accelerator School
Department of Physics
Massachusetts Institute of Technology
Boston, USA

Professor Victor Baryakhtar
ICSC World Laboratory, Branch Ukraine
WFS National Scholarship Programme
Kiev, the Ukraine

Professor Juan Manuel Borthagaray
Instituto Superior de Urbanismo
University of Buenos Aires
Buenos Aires, Argentina

Professor Enzo Boschi
President
National Institute for Geophysics and Vulcanology
Rome, Italy

Dr. Vladimir B. Britkov
Head of Information Systems Laboratory
Institute for System Analysis
Russian Academy of Sciences
Moscow, Russia

Dr. Franco Buonaguro
Director
Laboratory of Viral Oncology & AIDS Reference Center
Istituto Nazionale dei Tumori "Fondazione G. Pascale"
Co-Chairman WFS Panel on Medicine and Infectious Diseases
Naples, Italy

Dr. Gregory H. Canavan
Senior Fellow & Science Advisor
Los Alamos National Laboratory
Los Alamos, USA

Professor Georges Charpak
CERN
Geneva, Switzerland

Professor Luisa Cifarelli
Physics Department
University of Bologna
Bologna, Italy

Professor Robert Clark
Dept of Hydrology & Water Resources
College of Engineering and Mines, University of Arizona
Chairman WFS Panel on Extreme Weather Events
Tucson, Arizona

Professor Hassan Dalafi
International Centre for Theoretical Physics (ICTP)
Trieste, Italy (former)
United Nations Educational, Scientific and Cultural Organisation's (UNESCO)
Regional Office for Science and Technology in Europe (ROSTE)
Venice, Italy

Professor Guy de Thé
Institut Pasteur
Co-Chairman WFS Panel on Medicine and Infectious Diseases
Paris, France

Dr. Carmen Difiglio
Deputy Assistant Secretary for Policy Analysis
Office of Policy and International Affairs
US Department of Energy
Washington, D.C.

Colonel Mbareck Diop
Scientific Advisor to the President of Senegal
WFS National Scholarship Programme

Dr. Lorne G. Everett
Chancellor
Lakehead University
Chairman WFS Panel on Pollution
Ontario, Canada

Ambassador Francesco Paolo Fulci
Former Permanent Representative of Italy
to the United Nations and NATO

Dr. William Fulkerson
Joint Institute for Energy and Environment
University of Tennessee
Chairman WFS Panel on Energy
Knoxville, Tennessee

Dr. Bertil Galland
Writer and Historian
Buxy, France

Dr. Richard L. Garwin
IBM Fellow Emeritus
Thomas J. Watson Research Center, NY
Chairman WFS Panel on Mitigation of Terrorist Acts
New York, USA

Dr. Bernardino Ghetti
Indiana University Purdue University Indianapolis
Director Indiana Alzheimer Disease Center
Indiana University School of Medicine
Chairman WFS Panel on Brain and Behaviour
Indianapolis, USA

Professor Alberto González Pozo
Theory and Analysis Department
Universidad Autonoma Metropolitana
Xochimilco, Mexico

Sir Brian Heap
V.P. European Academies Science Advisory Council
The Royal Society London, England
and Honorary Fellow St Edmund's College Cambridge
Cambridge, UK

Dr. Walter F. Huebner
Southwest Research Institute
Chairman WFS Panel on Defence against Cosmic Objects
San Antonio, USA

Dr. Hisham Khatib
Honorary Vice Chairman - World Energy Council
Amman, Jordan

Dr. Vasiliy Krivokhizha
Deputy Chief
Council of Federation Staff International Department
Federal Assembly of the Russian Federation
Moscow, Russia

Professor Valery P. Kukhar
Director Institute for Bioorganic Chemistry and Petrochemistry
Ukrainian Academy of Sciences
President ICSC World Laboratory Ukraine Branch
Kiev, Ukraine

Professor Axel Lehmann
Institut fuer Technische Informatik
Universitaet der Bundeswehr Muenchen
Munich, Germany

Dr. Michael C. MacCracken
Chief Scientist for Climate Change Programs
Climate Institute
Washington, D.C.

Professor Bruno Maraviglia
Department of Physics
University of Rome "La Sapienza"
Rome, Italy

Professor Sergio Martellucci
Facolta di Ingeneria
Università degli Studi di Roma "Tor Vergata"
Rome, Italy

Dr. Léon M. Michaud
Italian Programme for the Rehabilitation of the
Meteorological Service of Iraq
Geneva, Switzerland

Professor Akira Miyahara
Emeritus Professor
National Institute for Fusion Science
Tokyo, Japan

Professor Povl Lebeck Ølgaard
Former Denmark Representative to the
NATO Scientific Committee

Dr. Jef Ongena
Ecole Royale Militaire
Laboratoire Plasmaphysics
Brussels, Belgium

Professor Gennady Palshin
ICSC – World Laboratory Branch Ukraine
Kiev, the Ukraine

Professor Andrei Piontkovsky
Strategic Studies Centre
Moscow, Russia

Professor Juras Požela
President ICSC - World Laboratory Branch Lithuania
Vilnius, Lithuania

Dr. Richard C. Ragaini
Department of Environmental Protection
Lawrence Livermore National Laboratories
Livermore, USA

Professor Zeononas Rokus Rudzikas
President of the Lithuanian Academy of Sciences
Vilnius, Lithuania

Professor Hiltmar Schubert
Director (former)
Fraunhofer-Institut für Chemische Technologie, ICT
Pfinztal, Germany

Professor Geraldo G. Serra
NUTAU São Paulo State University
São Paulo, Brazil

Professor Soroosh Sorooshian
Department of Civil & Environmental Engineering
and Earth System Science
University of California, Irvine
Irvine, USA

Dr. William A. Sprigg
Department of Atmospheric Sciences
Institute of Atmospheric Physics, University of Arizona
Chairman WFS Panel on Climate
Tucson, Arizona

Dr. Bruce Stram
Director
Element Markets
Houston, Texas

Professor Jan Szyszko
University of Warsaw
Warsaw, Poland

Professor Gerardus 't Hooft
Institute for Theoretical Physics
Universiteit Utrecht, and
Spinoza Institute
Utrecht, The Netherlands

Academician Evgeny Velikhov
Director
Russian Research Centre "Kurchatov Institute"
Moscow, Russia

Dr. Henning Wegener
Ambassador of Germany in Spain and delegate to NATO (former)
Chairman WFS Panel on Information Security
Berlin, Germany

Dr. Jody Westby
CEO
Global Cyber Risk LLC
Washington, D.C, USA

Professor Richard Wilson
Mallinckrodt Research Professor of Physics
Harvard University
Boston, USA

Dr. Lowell Wood
Hoover Institution
Stanford University
Stanford, USA